Food Packaging

ABOUT THE AUTHORS

Dr. (Mrs.) Neelam Khetarpaul presently working as Dean, College of Home Science at CCS Haryana Agricultural University, Hisar is the recipient of many awards *viz.*, Distinguished Teacher Award-2000, Young Investigator Award, Ms. Manju Utreja Gold Medal and cash award for doing the best research work in the University and Best Research Paper Award by AFST (I) Hisar Chapter. She is the recipient of various Visiting Fellowships abroad funded by different national and international agencies and visited USA, Australia, UK, Hungary and The Netherlands for academic pursuits. She is the country Coordinator of Indo-Netherlands Tailoring Biotechnologies project programme being funded by the Department of Biotechnology, Ministry of Science & Technology, Govt. of India and the Wageningen University, The Netherlands. She has published more than 180 research papers in various journals of national and international repute and 13 books in the discipline of Food Science and Human Nutrition.

Dr. Darshan Punia

Dr. Darshan Punia is presently working as Scientist in the Department of Foods and Nutrition. She was the Principal Investigator of National Agricultural Technology Project funded by ICAR. She is the recipient of ICAR Ch. Devi Lal Outstanding All India Coordinated Research Project (AICRP) Team Award, 2003. She also received Best Research Paper Presentation Award during the year 1998. She had been the convenor and co-convenor for poster presentations at International Conferences. She has published about 50 research papers in national and international journals of repute. Presently, she is the Incharge of All India Coordinated Research Project in the Department of Foods and Nutrition, CCS Haryana Agricultural University, Hisar.

Food Packaging

Prof. Neelam Khetarpaul
Ph.D. (Foods and Nutrition), FICN
Dean
College of Home Science
CCS Haryana Agricultural University
Hisar – 125 004 (Haryana), India
&
Dr. Darshan Punia
Scientist (Foods and Nutrition)
CCS Haryana Agricultural University
Hisar – 125 004 (Haryana), India

2012
DAYA PUBLISHING HOUSE
Delhi - 110 035

© 2012, PUBLISHER
ISBN 9789351240587

Published by   :   **Daya Publishing House**
                       A Division of
                       Astral International Pvt. Ltd.
                       – ISO 9001:2008 Certified Company –
                       4760-61/23, Ansari Road, Darya Ganj
                       New Delhi-110 002
                       Ph. 011-43549197, 23278134
                       E-mail: info@astralint.com
                       Website: www.astralint.com

Laser Typesetting :   **Classic Computer Services**
                       Delhi - 110 035

Printed at   :   **Chawla Offset Printers**
                       Delhi - 110 052

PRINTED IN INDIA

Preface

India is a major producer of agricultural products but it processes and packages just a fraction of the output. Packaging is at the heart of the marketing strategy in terms of functionality, shelf appeal and convenience. Packaging is becoming an essential part of the value chain analysis, regarding food safety, organoleptic characteristics, ergonomics and flexibility.

Life style changes are also happening at a fast pace. Traditional meals that "grandmothers cooked" are going out of fashion and people are increasingly opting for packed products. Convenience appears to be the single biggest driver in the ready meal market. Single person households, working women and those with higher disposable income have become increasingly influential in determining food production trends. Hence, packaging is of great importance in the final choice the consumer will make, because it directly involves convenience, appeal, information and branding.

In a very broad sense, the food industry is discovering the food service channel as a new distribution alternative. Alliances between food producers and food service operators will be the only means to develop successful solutions. The food industry has to "pull its packaging act together" if it hopes to exploit the international markets. The importance of packaging hardly needs stressing, since it is almost impossible to think of more than a handful of foods which are sold in an unpackaged state. Today packaging is essential and pevasive.

Food manufacturers know the importance of great food packaging. The food packaging could greatly affect the marketability of the product. But apart from serving as containers, food packaging also serves the purpose of opening up the avenues of communication between the manufacturer and the consumer. Food packaging helps make sure that the food is protected from tampering. It's also the means by which manufacturers transport and promote their goods. Apart from this, and probably the most important out of all the purposes of food packaging, is that it preserves the food and keeps it fresh for as long as possible. An important aspect of food packaging is **food packaging labelling.** The manufacturers are held accountable for any health hazards that the consumers may encounter because of poor packaging or poor food preservation techniques employed by the manufacturer. Furthermore, the fact that on average around 25 per cent of the ex-factory cost of consumer foods is for their packaging provides the incentive and the challenge for food packaging technologists to design and develop functional packages at minimum cost.

Food packaging is the major subject which draws on several disciplines; it is a multidisciplinary speciality. This book includes 18 chapters related to mechanical and chemical pulps, the kinds of deteriorative reactions, food packaging metals and their corrosion, packaging of foods in metal containers, use of glass in food packaging, thermoplastic polymers, important plastics processing methods, the packaging of cereals, dairy products, fruits and vegetables and meat and meat products, sterilization of packaging material and shelf life of packaged foods. Readers with an interest in food packaging will find the information given in various chapters to be timely, representative of some of the best work in the field of food packaging, and of great value.

We hope that this book shall be very useful for the students doing under graduation and post graduation in the disciplines of Food Science and Technology, Food Processing and Quality Control and Foods and Nutrition, research scholars and teachers. We shall regard our efforts amply rewarded if the book is received with enthusiasm.

We wish to express our sincere thanks to the publisher and especially to Mr. Anil Mittal for his continued interest in bringing out this edition.

Neelam Khetarpaul

Darshan Punia

Contents

Chapter 1

Introduction

Food packaging as an area of study draws on several disciplines, the major ones being chemistry, microbiology, food science and engineering.

Food packaging is an area with which, sooner or later, every practicing food scientist and technologist becomes involved. The importance of packaging hardly needs stressing, since it is almost impossible to think of more than a handful of foods which are sold in an unpackaged state. Furthermore, the fact that on average around 25 per cent of the ex-factory cost of consumer foods is for their packaging and provides the incentive and the challenge for food packaging technologists to design and develop functional packages at minimum cost.

In today's society, packaging is pervasive and essential. It surrounds, enhances and protects the goods we buy, from processing and manufacturing through handling and storage to the final consumer. Without packaging, materials handling would be a messy, inefficient and costly exercise, and modern consumer marketing would be virtually impossible.

The historical development of packaging has been well documented elsewhere and will not be repeated here. Suffice it to say that the highly sophisticated packaging industries which characterize modern society today are far removed from the simple packaging activities of earlier times. While an appreciation of the origin of packaging materials and a knowledge of the early efforts in the packaging arena are admirable, they are of little assistance to the modern packaging technologist and for this reason receive scant mention in this book, or indeed, in the standard reference on packaging.

Very few books can lay claim to be the first to expound or develop a particular area, and the present work is no exception. A number of books already exist with the words *packaging* or *food packaging* in their titles and this book builds on the pioneering efforts of these earlier authors. The whole field of food science and technology has undergone tremendous development over the last twenty years in particular, and this has been reflected in a plethora of books, many of which address quite specific subject areas.

Food packaging lies at the very heart of the modern food industry, and successful food packaging technologists must bring to their professional duties a wide-ranging background drawn from a multitude of disciplines. The inter-disciplinary nature of food packaging is evident from the chapter headings of this book. In the interests of balance, sufficient material has been included in the text for it to stand alone as a textbook for Honors year undergraduates and graduate students who are taking a two-semester course in food packaging. However, detailed bibliographies are included at the end so that those who wish to pursue particular aspects in more depth will have some references to start them on their way.

Those who already have an adequate background in food science and technology will find some of the chapters somewhat repetitive, while those with a background in engineering, while finding the chapters on packaging materials relatively familiar, will need to spend more time on those which focus on the major deteriorative reactions in foods and the principles underlying food preservation.

Despite the importance and key role which packaging plays, it is often regarded as a necessary evil or an unnecessary cost. Furthermore, in the view of many consumers packaging is, at best, somewhat superfluous, and, at worst, a serious waste of resources and an environmental menace. Such a viewpoint arises because the functions which packaging has to perform are either unknown or not considered in full. By the time most consumers come into contact with a package, its job in many cases is almost over, and it is perhaps understandable that the view that excessive packaging has been used, has gained some credence.

Packaging has been defined in a number of ways. A populist reference source defines packaging as an industrial and marketing technique for containing, protecting, identifying and facilitating the sale and distribution of agricultural, industrial and consumer products.

The Packaging Institute International defines packaging as the enclosure of products, items or packages in a wrapped pouch, bag, box, cup, tray, can, tube, bottle or other container form to perform one or more of the following functions: containment; protection and/or preservation; communications; and utility or performance. If the device or container performs one or more of these functions it is considered a package.

The UK Institute of Packaging provides three definitions of packaging:

1. A coordinated system of preparing goods for transport, distribution, storage, retailing and end-use;
2. A means of ensuring safe delivery to the ultimate consumer in sound condition at minimum cost;
3. A techno-economic function aimed at minimizing costs of delivery while maximizing sales (and hence profits).

It is important to distinguish between packaging as defined above, and packing which can be defined as the enclosing of an individual item (or several items) in a container, usually for shipping or delivery.

A distinction is usually made between the various "levels" of packaging. A primary package is one which is in direct contact with the contained product. It provides the initial and usually the major protective barrier. Examples of primary packages include metal cans, glass bottles, and plastic pouches. It is frequently only the primary package which the consumer purchases at retail outlets.

A secondary package contains a number of primary packages, *e.g.*, a corrugated case. It is the physical distribution carrier and is sometimes designed so that it can be used in retail outlets for the display of primary packages. A tertiary package is made up of a number of secondary packages, the most common example being a stretch-wrapped pallet of corrugated cases. In interstate and international trade, a quarternary package is frequently used to facilitate the handling of tertiary packages. This is generally a metal container up to 12 m in length which can hold many pallets and is intermodal in nature. That is, it can be transferred to or from ships, trains, and flatbed trucks by giant cranes. Certain designs are also able to have their temperature, humidity and gas atmosphere controlled and this is necessary in particular situations such as for the transportation of frozen foods or fresh fruits and vegetables.

Although the definitions given above cover in essence the basic role and form of packaging, it is necessary to discuss in more detail the functions of packaging and the environments where the package must perform those functions.

Now let us learn about the deterioration of packaged foods.

The principal aim of this chapter is to provide a brief overview of the major biochemical, chemical, physical and biological changes that occur in foods during processing and storage, and show how these combine to affect food quality. Knowledge of such changes is essential before a sensible choice of packaging materials can be made, since the rate and/or magnitude of such changes can often be minimized by selection of the correct packaging materials.

The deterioration of packaged foods (and this includes virtually all foods since today very few foods are sold without some form of packaging) depends largely on transfers that may occur between the internal environment inside the package, and the external environment which is exposed to the hazards of storage and distribution. For example, there may be transfer of moisture vapor from a humid atmosphere into a dried product, or transfer of an undesirable odor from the external atmosphere into a high fat product. In addition to the ability of packaging materials to protect and preserve foods by minimizing or preventing the transfers referred to, packaging materials must also protect the product from mechanical damage, and prevent or minimize misuse by consumers (including tampering).

Although certain types of deterioration will occur even if there is no transfer of mass (or heat, since some packaging materials can act as efficient insulators against fluctuations in ambient temperatures) between the package and its environment, it is possible in many instances to prolong the shelf life of the food through the use of packaging.

It is important that food packaging not be considered in isolation from food processing and preservation, or indeed from food marketing and distribution: all interact in a complex way, and concentrating on only one aspect to the detriment of the others is a surefire recipe for commercial failure.

The development of an analytical approach to food packaging is strongly recommended, and to achieve this successfully, the packaging technologist must have a good understanding of food safety and quality. The more important of these is without question food safety, that is the freedom from harmful chemical or microbial contaminants at the time of consumption. Packaging is directly related to food safety in two ways.

If the packaging material does not provide a suitable barrier around the food, microorganisms can contaminate the food and make it unsafe. However, microbial contamination can also arise if the packaging material permits the transfer of, for example, moisture or oxygen from the atmosphere into the package. In this situation, microorganisms present in the food but presenting no risk because of the initial absence of moisture or oxygen may then be able to grow and present a risk to the consumer.

Today aseptic packaging is the very common process.

Aseptic packaging can be defined as the filling of a commercially sterile product into sterile containers under aseptic conditions and sealing of the containers so that reinfection is prevented, *i.e.*, so that they are hermetically sealed. The term "aseptic" implies the absence or exclusion of any unwanted organisms from the product, package or other specific areas, while the term "hermetic" (strictly "air tight") is used to indicate suitable mechanical properties to exclude the entrance of microorganisms into a package and gas or water vapor into or from the package. The term "commercially sterile" has been defined as the absence of microorganisms capable of reproducing in the food under non-refrigerated conditions of storage and distribution, thus implying that the absolute absence of all microorganism need not be achieved.

Currently there are two specific fields of application or aseptic packaging:

1. Packaging of presterilized and sterile products, and
2. Packaging of a non-sterile product to avoid infection by microorganisms.

Examples of the first application include milk and dairy products, puddings, desserts, fruit and vegetable juices, soups, sauces, and products with particulates. Examples of the second application include fresh products such as fermented dairy products like yogurt.

There are three major reasons for the use of aseptic packaging:

1. To enable containers to be used that are unsuitable for in-package sterilization;
2. To take advantage of high-temperature-short-time (HTST) sterilization processes which are thermally efficient and generally give rise to products of a superior quality compared to those processed at lower temperatures for longer times, and
3. To extend the shelf life of products at normal temperatures by packaging them aseptically.

The first aseptic packaging of food (specifically milk in metal cans) was carried out in Denmark by Nielsen prior to 1913 and a patent for this process (termed aseptic conservation) was granted in 1921. In 1917, Dunkley in the USA patented sterilization with saturated steam of cans and lids and subsequent filling of a presterilized product. In 1923 aseptically packaged milk from South Africa reached a trade fair in London, England in perfect condition. The American Can Company developed a filling machine in 1933 called the Heat-Cool-Fill (HCF) system which used saturated steam under pressure to sterilize the cans and ends. The sterile cans were filled with sterile product and the ends sealed on in a closed chamber which was kept pressurized with steam or a mixture of steam and air. Three commercial plants were built and operated on this principle until 1945.

In the 1940s Martin developed a process in which empty metal cans were sterilized by treatment with superheated steam at 210°C prior to being filled with cold, sterile product. In 1950 the Dole Company bought the first commercial aseptic filling plant on the market.

At the end of the 1940s a dairy enterprise and machinery manufacturer in Switzerland (Alpura AG, Bern, and Sulzer AG, Winterthur, respectively) combined to develop UHT-sterilized, aseptically canned milk which was marketed in Switzerland in 1953. However, this system was not economical, mainly because of the cost of the cans, and Alpura, in collaboration with Tetra Pak AB of Sweden developed an aseptic cartoning system. Long shelf life milk packaged in this manner was first sold in Switzerland in October 1961.

Today packaging of horticultural products has become an important process.

Fresh fruits and vegetables are essential components of the human diet as they contain a number of nutritionally important compounds such as vitamins which cannot be synthesized by the human body. A fruit or vegetable is a living, respiring, edible tissue which has been detached from the parent plant. Fruits and vegetables are perishable products with active metabolism during the postharvest period. The shelf life of fruits and vegetables can be extended by, in simple terms, retarding the physiological, pathological and physical deteriorative processes (generally referred to as postharvest handling) or by inactivating the physiological processes (generally referred to as food preservation). Packaging has an important role to play in maximizing the shelf life of both the handling and preservation approaches to shelf life extension; this chapter will focus most attention on the former since the packaging requirements of the preservation processes are not unique to horticultural products.

Difficulties arise when attempts are made to draw a clear line between fruits and vegetables. Fruits and vegetables cannot be described clearly botanically or morphologically as they encompass numerous organs in vegetative or reproductive stages and belong to a large number of botanical families. Fruits tend to be restricted to reproductive organs arising from the development of floral tissues with or without fertilization. Vegetables, on the other hand, consist simply of edible plant tissues. Thus, some fruits are included as vegetables and *vice-versa*.

All fruits and vegetables continue to lose water through transpiration after they are harvested, and this loss of water is one of the main processes that affects their commercial and physiological deterioration. If transpiration is not retarded, it induces wilting, shrinkage, and loss of firmness, crispness and succulence, with concomitant deterioration in appearance, texture, and flavor. Most fruits and vegetables lose their freshness when the water loss is 3–10 per cent of their initial weight. As well as loss of weight and freshness, transpiration induces water stress which has been shown to accelerate senescence of fruits and vegetables.

Packaging of dairy products also the most important process is taking an important role in the preservation of dairy products.

For much of the world, particularly the West, milk from cattle (*Bas taurus*) accounts for nearly all the milk produced for human consumption. The composition of milk reflects the fact that it is the sole source of food for the very young mammal. Hence, it is composed of a complex mixture of lipids, proteins, carbohydrates, vitamins and minerals. The water phase carries some of the constituents in suspension while others are in solution. The fat is suspended in very small droplets as an oil in water emulsion and rises slowly to the surface on standing, a process often termed "creaming." For a detailed discussion of the chemistry of milk components, the reader is referred to a standard text.

Milk is processed into a variety of products, all having different and varying packaging requirements. The simplest product is pasteurized milk where, after a mild heat treatment, the milk is filled into a variety of packaging media and distributed. The shelf life of such a product varies from 2–10 days. UHT milk is subjected to a more complex process and the packaging is also more sophisticated; its shelf life can be up to 8 months. Cream is typically processed and packaged in a similar way to fluid milk and has a similar shelf life. Fermented dairy products, although being subjected to more complex processing operations, are also typically packaged in an analogous manner to fluid milk.

The other dairy products (butter, cheese and powders) are quite different in nature to fluid milk and their packaging requirements are therefore also quite different. Each of these groups of dairy products will be discussed in turn.

Milk for liquid consumption is often standardized with respect to fat content, and homogenized to retard the natural tendency for the fat globules to coalesce and rise to the surface. In response to consumer needs, a range of non-standard fluid milk products has been developed in recent years, such products having varying (reduced) fat levels (the term skim milk is used for such products) and additives such as calcium or other nutrients, *e.g.*, the fat-soluble vitamins. From a packaging point of view, these non-standard milks can be treated analogously to standard milks.

This function of packaging is so obvious as to be overlooked by many, but it is probably the basic function of packaging. With the exception of large, discrete products, all other products must be contained before they can be moved from one place to another. The "package," whether it be a milk bottle or a bulk cement rail wagon, must contain the product to function successfully. Without containment, pollution would be widespread.

The containment function of packaging make a huge contribution to protecting the environment from the myriad of products which are moved from one place to another on numerous occasions each day in any modern society. Faulty packaging (or underpackaging) would result in major pollution of the environment.

This is often regarded as the primary function of the package: to protect its contents from outside environmental effects, be they water, moisture vapor, gases, odors, microorganisms, dust, shocks, vibrations, compressive forces etc., and to protect the environment from the product. This is especially important for those products such as toxic chemicals which may seriously damage the environment.

In the case of the majority of food products, the protection afforded by the package is an essential part of the preservation process. For example, aseptically packaged milk and fruit juices in cartons only remain aseptic for as long as the package provides protection; vacuum-packaged meat will not achieve its desired shelf life if the package permits oxygen to enter. In general, once the integrity of the package is breached, the product is no longer preserved.

Packaging also protects or conserves much of the energy expended during the production and processing of the product. For example, to produce, transport, sell and store 1 kg of bread requires

15.8 mega joules (MJ) of energy. This energy is required in the form of transport fuel, heat, power and refrigeration in farming and milling the wheat, baking and retailing the bread, and in distributing both the raw materials and the finished product. To produce the polyethylene bag to package a 1 kg loaf of bread requires 1.4 MJ of energy. This means that each unit of energy in the packaging protects eleven units of energy in the product. While eliminating the packaging might save 1.4 MJ of energy, it would also lead to spoilage of the bread and a consequent loss of 15.8 MJ of energy.

Modern industrialized societies have brought about tremendous changes in life styles and the packaging industry has had to respond to those changes. One of the major changes has been in the nature of the family and the role of women. Now an ever-increasing number of households are single-person; many couples either delay having children or opt not to at all; there is a greater percentage than ever before of women in the work force.

All these changes, as well as other factors such as the trend towards "grazing" (*i.e.*, eating snack type meals frequently but on-the-run rather than regular meals), the demand for a wide variety of food and drink at outdoor functions such as sports events, and increased leisure time, have created a demand for greater convenience in household products: foods which are pre-prepared and can be cooked or reheated in a very short time, preferably without removing them from their primary package; condiments that can be applied simply through aerosol or pump action packages; dispensers for sauces or dressings which minimize mess, etc. Thus, packaging plays an important role in allowing products to be used conveniently.

Two other aspects of convenience are important in package design. One of these can best be described as the apportionment function of packaging. In this context, the package functions by reducing the output from industrial production to a manageable, desirable "consumer" size. Thus, a vat of wine is " apportioned" by filling into bottles; a churn of butter is "apportioned" by packaging into 25 gram minipats; a batch of ice cream is "apportioned" by filling into 2 liter plastic tubs.

Put simply, the large scale production of products which characterizes a modern society could not succeed without the apportionment function of packaging. The relative cheapness of consumer products is largely because of their production on an enormous scale and the associated savings which result. But as the scale of production has increased, so too has the need for effective methods of apportioning the product into consumer-sized dimensions.

An associated aspect is the shape (relative proportions) of the primary package in relation to convenience in use by consumers (*e.g.*, easy to hold, open and pour as appropriate) and efficiency in building into secondary and tertiary packages. In the movement of packaged goods in interstate and international trade, it is clearly inefficient to handle each primary package individually. Here packaging plays another very important role in permitting primary packages to be unitized into secondary packages (*e.g.*, placed inside a corrugated case) and then for these secondary packages to be unitized into a tertiary package (*e.g.*, a stretch-wrapped pallet). This unitizing activity can be carried a further stage to produce a quarternary package (*e.g.*, a container which is loaded with several pallets). If the dimensions of the primary and secondary packages are optimal, then the maximum space available on the pallet can be used. As a consequence of this unitizing function, materials handling is optimized since only a minimal number of discrete packages or loads need to be handled.

There is an old saying that "a package must protect what it sells and sell what it protects". It may be old, but it is still true; a package functions as a "silent salesman." The modern methods of consumer marketing would fail were it not for the messages communicated by the package. The ability of consumers to instantly recognize products through distinctive branding and labeling enables supermarkets to function on a self-service basis. Without this communication function (*i.e.*, if there were only plain packs and standard package sizes), the weekly shopping expedition to the supermarket would become a lengthy, frustrating nightmare as consumers attempted to make purchasing decisions without the numerous clues provided by the graphics and the distinctive shapes of the packaging.

Other communication functions of the package are equally important. Today the widespread use of modern scanning equipment at retail checkouts relies on all packages displaying a UPC (Universal Product Code) that can be read accurately and rapidly. Nutritional information on the outside of food packages has becomes mandatory in many countries.

But it is not only in the supermarket that the communication function of packaging is important. Warehouses and distribution centers would (and sometimes do) become symbolic if secondary and tertiary packages lacked labels or carried incomplete details. UPCs are also frequently used in warehouses where hand-held barcode readers linked to a computer make stock-taking quick and efficient. When international trade is involved and different languages are spoken, the use of unambiguous, readily understood symbols on the package is imperative.

Chapter 2

Mechanical and Chemical Pulps

2.1 Introduction

Pulp is the raw material for the production of paper, paperboard, corrugated board and similar manufactured products. It is obtained from plant fiber and it is a renewable resource.

The origin of papermaking, which is the formation of a cohesive sheet from the rebonding of separated fibers, has been attributed to Ts'ai-Lun of China in 105 AD, who used bamboo, mulberry bark and rags.

Since then many fibers have been used for the manufacture of paper including those from flax, bamboo and other grasses, various leaves, cottonseed hair and the woody fibers of trees. Today about 97 per cent of the world's paper and board is made from woodpulp, and about 85 per cent of the woodpulp used is from spruces, firs and pines-coniferous trees that predominate in the forests of the north temperate zone (brit). The influence of the raw material can be largely assigned to the length and wall thickness of the fibers rather than to their chemical composition. The main difference between low and high density woods is less paper-bonding of the latter at equal treatment.

Now let us study about the main constituents of the wood cell wall.

Cellulose

This is a long chain, linear polymer built up of a large number of glucose molecules (weight average degree of polymerization 8000–10,000 for native wood cellulose) and is the most abundant, naturally occurring organic compound. The fiber-forming properties of cellulose depend on the fact that it consists of long, relatively straight chains that tend to lie parallel to one another. Cellulose is moderately resistant to the action of chlorine and dilute sodium hydroxide under mild conditions, but is modified or dissolved under more severe conditions. It is relatively resistant to oxidation (*e.g.*, with bleaching agents) and therefore, bleaching operations can be used to remove small amounts of impurities such as lignin without appreciable damage to the strength of the pulp.

Hemicelluloses

These are lower molecular weight (weight average degree of polymerization 100–200) mixed-sugar polysaccharides consisting of one or more of the following molecules: xylose, mannose, arabinose, galactose and uronic acids, the composition differing from timber to timber. Hemicelluloses are usually soluble in dilute alkalis. The quantity rather than the chemical nature of the hemicelluloses appears to determine the paper properties. Hemi-celluloses are largely responsible for hydration and development of bonding during beating of chemical pulps.

Lignin

This is a highly branched, alkylaromatic, thermoplastic polymer of uncertain size, built up largely from substituted phenyl-propane units. It has no fiber-forming properties and is attacked by chlorine and sodium hydroxide with formation of soluble, dark brown derivatives. It softens at about 160°C.

The cell wall of softwoods, which are preferred for most pulp products, typically contain 40–45 per cent cellulose, 15–25 per cent hemicellulose and 26–30 per cent lignin by weight (brit). Compared to hardwoods, softwoods have fibers which are generally up to 2.5 times longer. As a result, hardwoods produce a finer, smoother but less strong sheet. The purpose of pulping is to separate the fibers without damaging them so that they can then be reformed into a paper sheet in the papermaking process. The intercellular substance (primarily lignin) must be softened or dissolved to free individual fibers. Commercial pulping methods take advantage of the differences between the properties of cellulose and lignin in order to separate fibers, but breaking and weakening of the fibers does occur to a greater or lesser degree at various stages during the pulping process.

Pulps which retain most of the wood lignin consist of stiff fibers that do not produce strong papers; they deteriorate in color and strength quite rapidly. These properties can be improved by removing most or all of the lignin by cooking wood with solutions of various chemicals, the pulps so produced being known as *chemical pulps.* In contrast, *mechanical pulps* are produced by pressing logs on to a grindstone, when the heat generated by friction softens the lignin so that the fibers separate with very little damage. Mechanical pulps can also be formed by grinding wood chips between two rotating refiner plates. In addition, there are some processes which are categorized as *semi-chemical* and *chemi-mechanical.*

Groundwood pulp is produced by forcing wood against a rapidly revolving grindstone. Practically all the wood fiber (both cellulose and lignin) is utilized compared to the several chemical processes where the lignin is dissolved to varying degrees. As a result, the yield of chemical pulp is about one half that of the mechanical process. The fibers vary in length and composition since they are in effect torn from the pulpwood.

Groundwood pulp contains a considerable proportion (70–80 per cent) of fiber bundles, broken fibers, and fines in addition to the individual fibers. The fibers are essentially wood with the original cell-wall lignin intact. Therefore, they are very stiff and bulky, and do not collapse like the chemical pulp fibers.

Most groundwood pulp is used in the manufacture of newsprint and magazine papers because of its low cost and quick ink absorbing properties (a consequence of the frayed and broken fibers). It is also used as board for folding and molded cartons, wallpapers, tissues and similar products. The paper has high bulk and excellent opacity, but relatively low mechanical strength.

In the 1950s the *refiner mechanical pulping* (RMP) process was developed which produced a stronger pulp and utilized various supplies of wood chips, sawmill residues and sawdust. However, the energy requirements of RMP are higher, and the pulp does not have the opacity of groundwood fibers.

Thermomechanical pulping (TMP) presteams chips to 110–150°C so that they become malleable and do not fracture readily under the impact of the refiner bars. This material is highly flexible and gives good bonding and surface smoothing to the paper. The production of TMP pulps increased dramatically after its introduction in the early 1970s because they could be substituted for conventional groundwood pulps in newsprint blends to give a stronger paper.

Chemithermomechanical pulping (CTMP) increases the strength properties of TMP pulps even further by a mild pretreatment with sodium sulfite. In general, CTMP pulps have a greater long-fiber fraction and lower fines fraction than comparable TMP pulps.

There are several chemical pulping methods, each based either directly or indirectly on the use of sodium hydroxide. Sufficient lignin is dissolved from the middle lamella to allow the fibers to separate with little, if any mechanical action. The nature of the pulping chemicals influences the properties of the residual lignin and the residual carbohydrates. For production of chemical pulps, the bark is removed and the logs passed through a chipper. The chipped wood is charged into a digester with the cooking chemicals, and the digestion carried out under pressure at the required temperature.

2.2 The Pulping Process

After pulping process the pulps vary considerably in their color and nature. Now let us study about sulfite processes.

The invention of this process is generally credited to the American chemist Benjamin Chew Tilghman who found in 1866 that cellulose fibers could be obtained by treating wood with a solution of bisulfite and sulfurous acid. A Swedish chemist, one C.D. Ekman, treated wood with magnesium bisulfite in 1870 and constructed the first sulfite paper mill in 1874 in Bergvik, Sweden. A

German modification of the American sulfite process was developed by the German chemist Mitscherlich and involved cooking (using indirect steam heating) at lower temperatures and pressures and for longer times than previously. The Austrians Eugen Ritter and Carl Kellner first practiced the direct injection of steam into a digester about 1878, reducing the time required for heating up the digester and thus, the total cooking time. Their procedure subsequently became known as the Ritter-Kellner process.

Several pulping processes are based on the use of sulfur dioxide as the essential component of the pulping liquor. Sulfur dioxide dissolves in water to form sulfurous acid, and a part of the acid is neutralized by a base in preparation of the pulping liquor. The various sulfite processes differ in the kind of base used and in the amount of base added, which governs the resulting acidity or pH of the liquor. These processes depend on the ability of sulfite solutions to render lignin partially soluble.

Acid Sulfite Process

This process (no longer of great commercial importance) uses an acidic liquor composed of calcium bisulfite and a considerable excess of sulfur dioxide. Wood cooked in this liquor gives a light-colored pulp which is easily bleached, most of the lignin and hemicelluloses dissolving in the liquor.

Bisulfite Process

This process uses a liquor composed of magnesium or ammonium bisulfite containing a slight excess of sulfur dioxide, resulting in a liquor which is not so strongly acidic. The delignification of wood chips takes place by cooking in bisulfite liquor at pH 4, the lignin reacting with the sulfurous acid to form soluble ligno-sulfonates which are later removed. The hemicelluloses are easily dissolved and removed accordingly.

The semichemical pulping idea was apparently first expressed by Mitscherlich in 1874 in a process involving softening of wood chips with sulfurous acid or bisulfite followed by rubbing or grinding to produce a pulp. It was started commercially through the development of a process to utilize certain by-product hardwood chips and then extended to other hardwoods which were used to make corrugated board.

The object of this process is to produce as high a yield as possible commensurate with the best possible strength and cleanliness. The first stage involves a mild chemical digestion using conventional type chemicals such as sodium hydroxide, sodium carbonate or sodium sulfate; the second stage involves mechanical disintegration. The hemicelluloses, mostly lost in conventional chemical digestion processes, are retained to a greater degree and result in an improvement in potential strength development. Semichemical pulps, although less flexible, resemble chemical pulps more than mechanical pulps.

The *neutral-sulfite semichemical* (NSSC) process makes use of sodium sulfite and a small amount of sodium hydroxide or sodium carbonate to give a slightly alkaline liquor. The NSSC pulp is obtained in higher yield but with a higher lignin content than in the other sulfite processes. It was developed specifically as a semichemical process for the production of corrugating medium, using unbleached pulp, and lends itself to small mills with minimal capital investment.

The digestion process consists essentially of the treatment of wood in chip form in a pressurized vessel under controlled conditions of time, liquor concentration and pressure/temperature. The main objects of digestion are:

1. To produce a well-cooked pulp, free from the non-cellulosic portions of the wood, *i.e.*, lignin and to a certain extent hemicelluloses.

2. To achieve a maximum yield of raw material, *i.e.*, wood pulp from pulp wood, commensurate with pulp quality.

3. To ensure a constant supply of pulp of the correct quality.

Today most pulping processes are continuous, and to give an indication of the processing conditions encountered, the widely used Kamyr kraft pulping process will be briefly described. After steaming at low pressure during which time turpentine and gases are vented to the condenser, the chips are brought to the digester pressure of 1000 kPa. They are picked up in a stream of pulping solution and their temperature raised to 170°C over 1.5 hours. After holding at this temperature for a further 1.5 hours, the digestion process is essentially complete.

After digestion, the liquor containing the soluble residue from the cook is washed out of the pulp which is then screened to remove knots and fiber bundles that have not fully disintegrated. The pulp is then sent to the bleach plant or paper mill.

Pulps vary considerably in their color after pulping, depending on the wood species, method of processing and extraneous components. The whiteness of pulp is measured by its ability to reflect monochromatic light in comparison to a known standard (usually magnesium oxide). Cellulose and hemicellulose are inherently white and do not contribute to color; it is the chromophoric groups on the lignin that are largely responsible for the color of the pulp.

Table 2.1: Approximate Brightness Ranges of Various Unbleached Pulps

Brightness Range	Type of Pulp
15–30	Kraft
40–50	NSSC, Bisulfite, Soda
50–60	Groundwood, Sulfite

Basically there are two types of bleaching operations: those that chemically modify the chromophoric groups by oxidation or reduction but remove very little lignin or other substances from the fibers, and those that complete the delignification process and remove some carbohydrate material.

Chemical methods must be used to improve the color and appearance of the pulp; these are bleaching treatments and involve both the oxidation of colored bodies and the removal of residual encrusting materials (the principal one being lignin) remaining from the digestion and washing stages. Since bleaching reduces the strength of the pulp, it is necessary to reach a compromise between the brightness of the finished sheet and its tensile properties.

The most effective bleaching agent for most groundwoods is hydrogen peroxide, and since the bleaching is performed in alkaline solutions, sodium peroxide is also used. The reaction requires 3 hours at 40°C and is followed by neutralization and destruction of excess peroxide with SO_2. These pulps may be improved in color to only a limited extent since they contain virtually all the lignin from

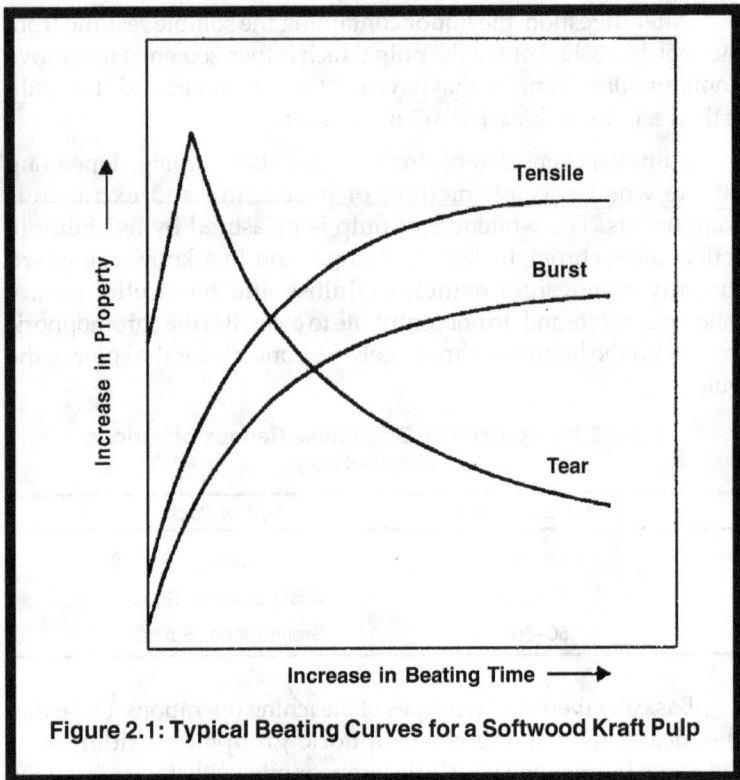

Figure 2.1: Typical Beating Curves for a Softwood Kraft Pulp

the original wood. Peroxide bleaching allows brightness to be increased by nearly 20 per cent.

Chemical pulps vary a little when while compared to mechanical pulps.

Here the reagents for full bleaching are mostly oxidative, and since the carbohydrates are also susceptible to oxidation, bleaching is accomplished under the mildest conditions. Bleaching of chemical pulps is basically stepwise purification of colloidal cellulose, and thus, bleaching can be regarded as a continuation of the cooking process. The bleaching of pulp is done through chemical reactions of bleaching agents with lignin and coloring matter of the pulp. The bleaching is performed in a number of stages utilizing chlorine, hypochlorite, chlorine dioxide, oxygen and peroxide. Between these stages, the pulp is treated with alkali to dissolve degradation products.

Chlorine gas and chlorine dioxide are the major bleaching chemicals used, and concerns have been expressed about the formation of trace levels of dioxins as a result of the bleaching process. A relatively new development which is now fairly common in the industry is the replacement of conventional bleaching agents such as chlorine, hypochlorites and chlorine dioxide with oxygen, ozone and peroxides; details of these processes have been given elsewhere. These changes have been introduced to enable pulp and paper mills to meet tough new anti-pollution laws and regulations and to conserve wood, chemicals and energy.

Stock preparation is defined as that part of the pulp and papermaking process in which pulp is treated mechanically and, in some instances, chemically by the use of additives, and is thus, made ready for forming into a sheet or board on the paper machine. During the stock preparation steps, the pulps are most conveniently handled as aqueous slurries. However, in the papermaking process, utilizing purchased pulps and waste paper which are received as dry sheets, the first step is the separation of all the fibers from one another, and their dispersion in water with a minimum of mechanical work to avoid altering the fiber properties. This process is known as slushing and is carried out on machines such as the Hydrapulper or the Sydrapulper. When the pulping and papermaking operations are adjacent to one another, pulps are usually delivered to the paper mill in slush form directly from the pulping operation.

Now, let us study about stock preparation methods. There are two main methods which are in practice *i.e.,*

1. Beating
2. Refining.

The object of beating is to increase the surface area of the fibers by assisting them to imbibe water. As a result, additional bonding opportunities are provided for between cellulose molecules of neighboring fibers. The beating also makes the fibers more flexible, causing them to become relatively mobile and to deform plastically on the paper machine.

The mixture of pulp (known as the furnish) is passed into the beater and brought to a consistency of 5–7 per cent. The fibers are then beaten while suspended in the water in order to impart to them many of the properties that will determine the character of the final product.

The quality and characteristics of the finished paper depend to a great extent on the treatment in the beater. Because papermaking fibers are stiff and springy, the resulting paper would be flabby and weak if made into a sheet of paper without beating. There would be little adhesion between the fibers and they could not be consolidated under the presses of the papermachine. A sheet formed from an unbeaten pulp has a low density, and is rather soft and weak, whereas if the same pulp is beaten, the resultant paper is much more dense, hard and strong. If taken to the extreme, beating produces very dense, translucent and glassine-type sheets. Thus, beating can be controlled to produce papers as widely different as blotting and greaseproof.

A small amount of beating will produce a highly absorbent sheet with high tear strength but low burst and tensile strength. As beating progresses, burst and tensile strength values rise to a maximum, followed by a fairly rapid fall as the fibers become shortened and lose their structure. The bulk decreases and the density increases, opacity is lowered and transparency increased. The tensile strength, bursting strength and tearing resistance are shown as functions of beating time for a softwood kraft pulp.

Since its invention in Holland around 1690, the principle of the batch-operated hollander beater has remained substantially the same. It consists of a cylindrical roll containing knives which revolves over a stationary bedplate which also contains a set of knives. Circulating stock passes between the roll and the bedplate, the severity of beating being controlled by adjusting the load of one against the other. Circulation is continued until the pulp is considered ready to be made into the desired paper.

In many papermills, beaters have been replaced by continuous refiners, of which the disk refiner (rotary disks rotate against a working surface) is the most recent development. However, the batch beater is a convenient vessel for adding chemicals and mixing them intimately with the pulp in order to give special properties to the final papers.

In papermaking, chemicals such as strength additives, adhesives, mineral fillers and sizing agents may be added at the beater stage prior to sheet formation (*i.e.*, internal addition), or to the resulting sheet after complete or partial drying, depending primarily on the desired effects. Strength additives usually are added internally if uniform strength throughout the sheet is wanted, but they are applied to the surface if increased surface strength is needed. Fillers can improve brightness, opacity, softness, smoothness and ink receptivity, and are essentially insoluble in water under the conditions of use. The main drawback is that the materials added may be lost through the wire of the papermachine in the large amount of water used. Therefore, if an additive cannot be retained efficiently from a dilute pulp slurry, it is better to apply it to the surface of the sheet.

Sizing is the process of adding materials to the paper in order to render the sheet more resistant to penetration by liquids, particularly water. Rosin is the most widely used sizing agent, but starches, glues, caseins, synthetic resins and cellulose derivatives are also used. The sizing agents may be added directly to the stock as beater additives to produce internal or engine sizing, or the dry sheet may be passed through a size solution to produce a surface size.

2.3 Paper Conversion and Surface Treatments

Almost all paper is converted by undergoing further treatment after manufacture, such as embossing, impregnating, saturating, laminating and the forming of special shapes and sizes such as bags and boxes. Further surface treatment involving the application of adhesives, functional products and pigments are common, depending on the end use of the paper. Surface treatment applies similar materials on the web.

Because of the widespread use of paper and paperboard in direct contact with foods, most mills require paper chemicals that have been cleared for use with food by regulatory authorities such as the FDA. It is recommended that the regulatory status of any paper-chemical additive be determined with the supplier prior to use.

Surface treatments such as sizing and coating are extensively applied to improve the appearance of products. Paper may be coated either on equipment that is an integral part of the paper machine (*i.e.*, on-machine coating), or on separate converting equipment. The most common method for the application of chemicals to the surface of a paper web is by a size press. In the size press, dry paper is passed through a flooded nip and a solution or dispersion of the functional chemical contacts both sides of the paper. Excess liquid is squeezed out in the press and the paper is redried.

Surface-sizing agents prevent excess water penetration and improve the strength of the paper. The sizing agent penetrates far enough into the paper to increase the fiber bonding and the dependent properties such as bursting, tensile and folding strengths. An additional effect is an improvement in the scuffing resistance of the paper surface.

The rate of flow of a liquid through a very thin tube or capillary can be represented by the Washburn equation which combines the equation representing the natural driving force for fluid movement in a capillary tube with the Poiseuille equation for laminar flow through a tube. Two of the five parameters that govern the rate of flow, *viz.*, surface tension and viscosity of the penetrating liquid, depend on the use to which the sized paper product is put. Two more, *viz.*, the radius and length of the pores, are governed by the papermaker's products, *i.e.*, the basis weight, bulk density and porosity of the sheet. It is obviously important that the appropriate chemical is selected to produce the sizing that is required by the paper producer and the user.

The most commonly used materials for surface sizing are starches, usually chemically modified *e.g.*, oxidized starches, cationic starches and hydroxyethylated derivatives. Other film-forming materials which can be used for surface sizing include carboxymethyl cellulose and poly (vinyl alcohol) which provide oil- and grease-repellent coatings and improve paper strength. Other polymeric sizing agents such as polyurethanes are used as surface-sizing and strength-enhancing agents. Their cost is relatively high compared with other sizing agents.

Fluorochemical emulsion sizing agents can be applied to the surface of paper or paperboard to provide good oil and grease repellancy. They find application for pet-food bag papers, meat, fish and poultry wrap, cookie bags and candy wrappers.

2.4 Pigments and Adhesives

Pigments comprise 70–90 per cent of the dry solids in paper coatings and are generally designed to mask or change the appearance of the base stock, improve opacity, impart a smooth and receptive surface for printing, or developing a practical paper machine, eventually becoming bankrupt and dying in poverty. Despite their misfortune, their name has been familiar to generations of papermakers for the development of a machine, the essential principles of which are still in use today.

Paper is made by depositing a very dilute suspension of fibers from a very low consistency aqueous suspension (greater than 99 per cent water) on to a relatively fine woven screen, over 95 per cent of the water being removed by drainage through the wire. The fibers interlace in a generally random manner as they are deposited on the wire and become part of the filter medium.

Although paper is the general term for a wide range of matted or felted webs of vegetable fiber that have been formed on a screen from a water suspension, it is usually subdivided into paper and paperboard. However, there is no rigid line of demarcation between the two, with board often being defined as stiff and thick paper. ISO standards define paperboard as paper with a basis weight (grammage) generally above 250 grams per square meter, but there are exceptions.

The modern fourdrinier papermachine consists essentially of an endless woven wire gauze or forming fabric stretched over rollers. The suspension of pulp in water is forced on to the wire in a regular stream, the relative speeds of the stock and wire affecting the degree to which the fibers are aligned along the direction of travel. As it progresses down the wire, the fiber slurry loses water, first by free drainage and then under vacuum.

Fourdriniers are standard in the industry and are used to produce all grades of paper and paperboard. Heavy paperboards require a long drying time and machine speeds are 10–250 meters per minute. Very dense papers (*e.g.*, glassine and greaseproof) are difficult to dewater in the forming and press sections and speeds range from 20 to 300 meters per minute, depending on the product. Brown grades (*e.g.*, paper bags and linerboard) are produced at 200–800 meters per minute depending on basis weight. The majority of newsprint machines operate at 600–900 meters per minute.

We have studied about pigments. Now let us have a birdview on functions of adhesives.

The primary function of the adhesive in pigment coating is to bind the pigment particles together and to the raw stock. The type and proportion of the adhesive controls many of the characteristics of the finished paper such as surface strength, gloss, brightness, opacity, smoothness, ink receptivity and firmness of the surface. The strength must be sufficient to prevent the coating from being picked up by tacky printing inks.

Animal glue was the first material used for bonding paper and as an adhesive in paper coating, but it has been replaced by casein and soy protein. Starches are used in many coated papers, and are the principal adhesives if resistance to moisture is not required. Various rubber latexes and other emulsions are also employed as adhesives. Acrylic-based emulsions are used mostly on paperboard, and their freedom from odor makes them ideally suited for use in food packaging.

In many packaging applications, a barrier may be needed against water vapor or gases such as oxygen. A water barrier can be formed by changing the wettability of the paper surface with sizing agents. Coating the paper with a continuous film of a suitable material will confer gas or vapor barrier properties. Paraffin wax applied in a molten form was commonly used to give a water vapor barrier, but polyethylene applied by extrusion gives a more durable and flexible coating. Highly sophisticated coatings based on a wide range of polymers and modifiers can be formulated using solvent systems. The polymers include cellulose derivatives, rubber derivatives, vinyl copolymers, polyamides, polyesters and butadiene-styrene copolymers.

2.5 Fine Papers and Coarse Papers

Paper is mainly divided into two main types:

1. Fine Papers
2. Coarse Papers

Fine papers, generally made of bleached pulp, and typically used for writing paper, bond, ledger, book and cover papers, and coarse papers, generally made of unbleached kraft softwood pulps and used for packaging.

Kraft Paper

This is typically a coarse paper with exceptional strength, often made on a fourdrinier machine and then either machine-glazed on a Yankee dryer or machine-finished on a calender. It is sometimes made with no calendering so that when it is converted into bags, the rough surface will prevent them from sliding over one another when stacked on pallets.

Bleached Paper

These are manufactured from pulps which are relatively white, bright and soft and receptive to the special chemicals necessary to develop many functional properties. They are generally more expensive and weaker than unbleached papers. Their aesthetic appeal is frequently augmented by clay coating on one or both sides.

Greaseproof Paper

This is a translucent, machine-finished paper which has been hydrated to give oil and grease resistance. Prolonged beating or mechanical refining is used to fibrillate and break the cellulose fibers which absorb so much water that they become superficially gelatinized and sticky. This physical phenomenon is called hydration and results in consolidation of the web in the papermachine with many of the interstitial spaces filled in.

The satisfactory performance of greaseproof papers depends on the extent to which the pores have been closed. Provided that there are few interconnecting pores between the fibers, the passage of liquids is difficult. However, they are not strictly "greaseproof" since oils and fats will penetrate them after a sufficient interval of time. Despite this, they are often used for packaging butter and similar fatty foods since they resist the penetration of fat for a reasonable period.

Waxed Paper

Waxed papers provide a barrier against penetration of liquids and vapors. A great many base papers are suitable for waxing, including greaseproof and glassine papers. The major types are wet-waxed, dry-waxed and wax-laminated. Wax-sized papers, in which the wax is added at the beater during the papermaking process, have the least amount of wax and therefore, give the least amount of protection.

Table 2.2: Main Types of Packaging Papers

Basic Material	Source	Weight Range Kg/1000 m²	Tensile Strength kg/m	Properties and Uses
Kraft papers	Sulfate pulp from softwoods	70–300	250–1150	Heavy-duty paper, bleached, natural or colored; may be wet-strengthened or made water-repellent. Used for bags, multi wall sacks and liners for corrugated board. Bleached varieties are used for food packaging where strength is required.
Sulfite papers	Usually bleached generally made from mixture of softwood and hardwood	35–300	Very variable	Clean, bright paper of excellent printing nature and used for smaller bags, pouches, envelopes, waxed papers, labels and for foil laminating, etc.
Greaseproof papers	From heavily beaten pulp	70–150	180–450	Grease-resistant for baked goods and fatty foods
Glassine papers	Similar to grease-proof but super-calendered	40–150	140–535	Oil- and grease-resistant, used as an odor barrier for lining bags, boxes, etc. and for greasy foods
Vegetable parchment	Treatment of unsized paper with concen-trated sulfuric acid	12–75	215–1450	Non-toxic, high wet strength, grease- and oil-resistant for wet and greasy food
Tissue papers	Lightweight paper from most pulps	20–50	Low	Lightweight, soft wrapping paper

Wet-waxed papers have a continuous surface film on one or both sides, achieved by shock-chilling the waxed web immediately after application of the wax. This also imparts a high degree of gloss on the coated surface. Dry-waxed papers are produced using heated rolls and do not have a continuous film on the surfaces. Consequently, exposed fibers act as wicks and transport moisture into the paper. Wax-laminated papers are bonded with a continuous film of wax which acts as an adhesive. The primary purpose of the wax is to provide a moisture barrier and a heat sealable laminant.

Often special resins or plastic polymers are added to the wax to improve adhesion and low temperature performance, and to prevent cracking as a result of folding and bending of the paper.

Glassine Paper

Glassine paper derives its name from its glassy, smooth surface, high density and transparency. It is produced by further treating greaseproof paper in a supercalender where it is carefully dampened with water and run through a battery of steam-heated rollers. This results in such intimate interfiber hydrogen bonding that the refractive index of the glassine paper approaches the 1.02 value of amorphous cellulose, indicating that very few pores or other fiber/air interfaces exist for scattering light or allowing liquid penetration. The transparency can vary widely depending on the degree of hydration of the pulp and the basis weight of the paper. The addition of titanium dioxide makes the paper opaque, and it is frequently plasticized to increase its toughness.

Vegetable Parchment

Vegetable parchment takes its name from its physical similarity to animal parchment (vellum) which is made from animal skins. The process for producing parchment paper was developed in the 1850s, and involves passing a web of high-quality, unsized chemical pulp through a bath of concentrated sulfuric acid. The cellulosic fibers swell and partially dissolve, filling the interstices between the fibers and resulting in extensive hydrogen bonding. Thorough washing in water, followed by drying on conventional papermaking dryers, causes reprecipitation and consolidation of the network, resulting in a paper that is stronger wet than dry (it has excellent wet strength, even in boiling water), free of lint, odor and taste, and resistant to grease and oils. Unless specially coated or of a heavy weight, it is not a good barrier for gases.

Because of its grease resistance and wet strength, it strips away easily from food material without defibering, thus finding use as an interleaver between slices of food such as meat or pastry. Labels and inserts in products with high oil or grease content are frequently made from parchment. It can be treated with mold inhibitors and used to wrap foods such as cheese.

Parchment paper with great shock-absorbing capability can be produced by wet creping, resulting in extensibility combined with natural tensile toughness. Special finishing processes provide qualities ranging from rough to smooth, brittle to soft, and sticky to releasable. It was first used for wrapping fatty foods such as butter, and still finds such an application today.

Glazed imitation parchment (GIP) is made from strong sulfite pulp which is heavily engine-sized and glazed to give the necessary degree of protection.

Tissue Paper

Tissue papers range from semitransparent to totally opaque, and can be waxed. They are generally either machine-finished (MF) or machine-glazed (MG), the former process involving calendering between highly-polished rollers which squeeze and polish both sides of the paper, and the latter process producing a smooth glazed side and a rough back side. MG papers may also be machine finished to improve the smoothness on both sides.

2.6 The Uses of Papers

A variety of papers are used in the manufacture of paper bags depending on the end use of the bags. Both bleached and unbleached papers have been used, but the most common are the brown paper bags made from kraft paper. Bags can differ in shape, style and the number of piles, *e.g.*, single, double and multiwall. Multiply (or multiwall) paper bags are also referred to as paper sacks.

Single-wall bags can be simple flat bags with a lengthwise seam and a base where the paper is folded under the glued; satchel-bottom bags which provide a flat base when filled; and square-bottom bags which are similar to flat-bottom bags except that they contain bellows folds at the sides to reduce the width of the closed bag without reducing capacity.

Multiwall bags are used primarily for the packaging of materials such as powders or granules that need no protection against compressive forces, since their principal function is to absorb energy without rupturing, contain the contents and protect them from contamination. It is important that the paper has strength when under load, and that it should stretch under these conditions. Therefore, tensile strength and stretch characteristics of the paper are important. The strength of the paper (and therefore, the performance of the bags made from it) varies with the moisture content of the paper. Dry conditions tend to lower the strength of the bag, while the strength increases with moisture content up to a certain level.

Multiwall bags contain from two to six plies of high quality paper, sometimes with special coatings such as polyethylene to provide a liquid water and water vapor barrier. Each ply is fabricated as a tube and arranged one within the other so that each layer bears its share of any applied or induced stress. Generally better performance is obtained against shock or impact by the use of several plies of relatively lightweight papers rather than by the use of fewer plies of heavier-weight papers, although the latter is considered to be more effective against externally applied point stresses such as protruding nails or sharp corners on pallets.

The average heavy-duty multiwall bag is constructed of a number of plies of paper with basis weights of 65–114 gsm, the most frequently used basis weights being 65, 81 and 98 gsm.

Paper is generally termed board when its substance exceeds 250 gsm, except in the United Kingdom where the dividing line is arbitrarily taken as 220 gsm.

A variety of types of paperboards are manufactured. The simplest types are single-ply thick papers made from 100 per cent bleached chemical wood pulp; they are used in food packaging where purity and clean appearance is required together with a degree of strength.

The methods of fiber treatment are essentially the same as those used in manufacturing paper grades. After the sheet has been formed, the methods used for removing excess water and finishing the web are also essentially the same as those used in the manufacture of paper. The main difference is found at the forming section of the machine where the web is formed.

Rigid paper containers are generally constructed of paperboard or a combination of paperboard and paper.

Folding Cartons

Folding cartons are containers made from sheets of paperboard (typically with thicknesses between 300 µm and 1100 µm) which have been cut and scored for bending into desired shapes; they are delivered in a collapsed state for erection at the packaging point.

The boards used for cartons have a ply structure and many different structures are possible, ranging from recycled fibers from a variety of sources (as in chipboard), through fibers where the outer ply is replaced with better quality pulps to give white-lined chipboards, to duplex boards without any waste pulp and solid white boards made entirely from bleached chemical pulp. The latter are used for frozen food packaging and other applications where the board must be waxed.

A number of steps are involved in converting paperboard into cartons. Where special barrier properties are required, coating and laminating is carried out; wax lamination provides a moisture barrier, lining with glassine provides grease resistance, and laminating or extrusion coating with plastic materials confers special properties including heat sealing. The use of barrier materials in cartonboard is restricted by the inability of the normal types of carton closure to prevent the ingress of moisture directly. Consequently, considerable effort has been directed towards the development of one-piece coated cartons which have improved barrier properties. Liquid-tight, hermetically-sealed cartons are now widely used for the packaging of a wide range of liquid foods, such packaging being able to be carried out under aseptic conditions to give packs which will retain the product in a commercially sterile state. Incorporation of a liner which can be sealed into a bag can give a carton able to contain a liquid, or give gas and water vapor protection to the contents, and such systems are available. Other developments have included the incorporation of various types of tear-off, easy-open and dispensing devices into cartons.

Coating of the outer board greatly enhances the external appearance and printing quality, and clay and other minerals are used for such purposes. The coating can be applied either during the board-making operation or subsequently. Foil-lined boards are also used for various types of carton, not only to provide barrier properties (as in aseptic cartons) but also (in certain applications) to improve reheatability of the contents, especially in microwave ovens.

The conventional methods of carton manufacture involve printing of the board, followed by creasing and cutting to permit the subsequent folding to shape, the stripping of any waste material which is not required in the final construction, and the finishing operation of joining appropriate parts of the board, either by gluing or (occasionally) stitching. During creasing and folding, cartonboard is subjected to complex stresses, and the ability of a board to make a good carton depends on its rigidity, ease of ply delamination, and the stretch properties of the printed liner. It is important that the surface layer on the top of the board is of an elastic nature and relatively high strength compared with the properties of the underlying layers since they will be in compression.

2.7 The Construction of Corrugated Board

Corrugated board is manufactured on a very large and specialized machine (really a series of machines) called a corrugator which can be divided into three sections. In the first (known as the single facer), the medium (a relatively thin web of paperboard) is conditioned with heat and steam and then passed between two fluted metal rolls to form the corrugation. Starch adhesive is applied to the tips of the flutes, and the conditioned liner is brought into contact with the flute tips to form the adhesive bond.

In the second section the single-face web is preheated, adhesive applied to the flutes and the double-backer liner joined to the single-face web. After cooling, the board enters the third section where it is trimmed using a slitter and then scored to define the flaps and depth of the box, a process which involves crushing the flutes in a narrow groove in the machine direction so that the box can be folded along these lines later when it is assembled. The board is then cut to predetermined lengths.

Double-wall corrugated board is made by combining two fluted corrugating mediums with a central liner and then adding two outer liner facings. Sometimes three corrugated mediums with two inner liners and two outer facings are combined to form triple-wall corrugated board.

After leaving the corrugator the blanks pass to a printer-slotter. Here they are printed using soft rubber dies, and body scores and slots are made. The body scores are similar to the flap scores made in the corrugator except that they are parallel to the flutes, thereby forming the vertical edges of the box when it is assembled. The slots permit the side flaps to fold over the end flaps, the flaps forming the top and bottom of the box. When the printing and slotting operations have been completed, the blanks are formed into a flat tube by a variety of methods: gluing the two ends of the box; stitching with metal staples; or taping with paper or cloth tape. The flat tubes are then palletized for shipment.

A number of variations to the above procedures are possible: different weight facings and different flutes can be used. As well, innovative designs have been developed to produce boxes which can be set up by the packager without the need for additional gluing or stapling to form special purpose containers.

An international code (known as the FEFCO/ASSCO Fiberboard Case Code) has been agreed on between the various

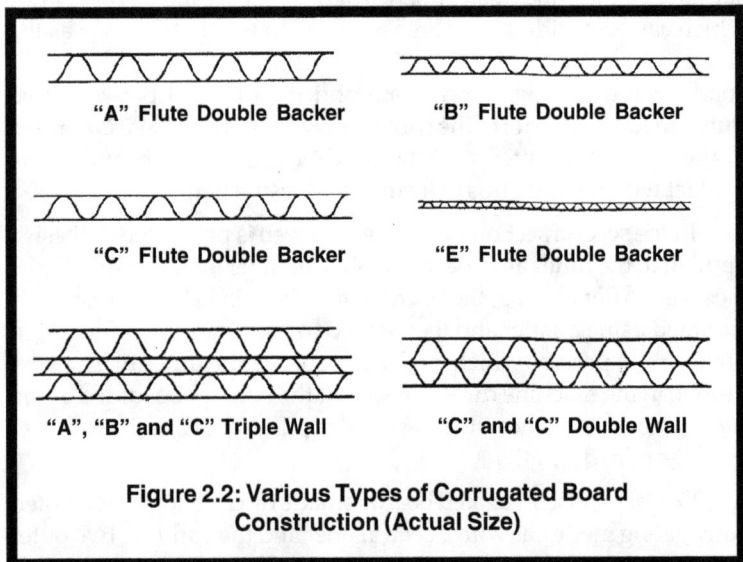

Figure 2.2: Various Types of Corrugated Board Construction (Actual Size)

fiberboard case-making organizations of many countries, and it is widely used. Although there are a wide variety of styles of corrugated fiberboard containers, the most popular is the regular slotted container (RSC), due mainly to its simple construction and economical board usage in terms of board area to volume ratio of the fabricated container. However, its poor stacking strength is its major disadvantage, *i.e.,* it rates poorly in terms of volume over strength of fabricated containers. Therefore, when the product can be relied upon to carry most of the static and dynamic stacking load forces (as is the case for canned foods), or when internal partitions can carry a significant proportion, the economy of the RSC is unbeatable. However, when the container has to provide protection in addition to simple containment, other less economical container styles must be resorted to.

Because there are a number of products where it is undesirable for them to carry a significant part of the stacking loads (*e.g.,* fresh produce), there is a continual program of research and development to devise new container styles which will use less board and provide greater stacking strength. Since most of the stacking load is carried by the corners of the container, increasing the wall thickness at the corners makes most efficient use of the material.

The use of mathematical models to aid the design of corrugated containers has become increasing prevalent over the last 20 years. Two basic approaches are possible in constructing a mathematical model to predict corrugated container stacking strength: the properties of the material (*e.g.,* paperboard facings and corrugating medium) are correlated to the stacking strength of the fabricated container; or the container is viewed as a basic structure, *e.g.,* single degree of freedom system with inherent viscoelastic properties.

The former is the traditional approach and was used by McKee who adapted a well-known semiempirical formula commonly used in prediction of failure in shell-type structures for prediction of maximal top-to-bottom strength of corrugated shipping containers. The ultimate compressive strength of a RSC can be related to the board caliper, container perimeter and edgewise compressive strength of the corrugated paperboard by a simplified version of this formula. This enables a direct link to be established between the corrugated paperboard manufacturing process and the ultimate strength of the finished container, providing an invaluable tool for paper manufacturers and corrugated paperboard converters, many of whom have developed in-house computer-aided design systems to assist in container development. However, the end user of the container is primarily interested in the performance characteristics of the container in the field, irrespective of the type of paper it is made from, and ultimate static compressive strength in ideal environmental conditions is only one of several indices of container performance as Peleg has shown.

2.8 Solid Fiberboard

Solid fiberboard consists of numerous bonded plies (typically two to five, with three- and four-ply being the most common) of container board lined on one or both faces with kraft or similar paper between 0.13 and 0.30 mm thick to form a solid board of high strength. The total caliper of the lined board ranges from 0.80 to 2.8 mm. Being solid, it is consequently much heavier in weight for a given thickness than corrugated board, the combined weight of the component plies ranging from 556–1758 gsm.

Solid fiberboard containers are generally two to three times the cost of comparable corrugated containers and are therefore, used almost exclusively for applications in which container return and reuse are possible. Solid fiberboard containers can be re-used satisfactorily from 10 to 15 times.

Solid fiberboard can be made by passing two or more webs or plies of paperboard between a number of sets of press rolls, adhesive being applied to each ply before it passes through the press nips. In multiple structures it is usual to use a poor grade of paperboard (*e.g.*, chipboard) in the central plies and a strong linerboard as the outside facings or liners [baum]. After formation of the solid fiberboard, the subsequent operations are similar to those described for corrugated boxes. Many variations of style, quality and properties are possible in both materials.

Now let us study about the common tests used in the manufacture of corrugated board.

The more common tests performed on linerboard used in the manufacture of corrugated board are: bursting strength, basis weight, puncture, stiffness, tear, tensile, moisture, caliper, water absorbency, pH, compression resistance and water drop. There are appropriate ASTM, BS and TAPPI standards for these tests, but the more important ones will be briefly mentioned here.

Bursting strength uses a Mullen tester in which the sample is clamped between two concentric platens which each have a circular opening in the center. A rubber diaphragm under the bottom of the sample is expanded by hydraulic pressure at a controlled rate until the sample ruptures; the reading on a pressure gauge at this point gives the bursting strength value in force per unit area. This is an important control test in the paper mill or corrugating plant because the bursting strength of the linerboard I has an important influence over the finished container properties.

However, of more importance than bursting strength in determining the end-use performance of the finished corrugated container is compressive strength. A measure of the edgewise resistance of corrugating medium to compression is given by the ring crush or stiffness test which is used to indicate edgewise rigidity of the board and the probable crushing resistance of the finished container. The test essentially involves crushing a ring specimen rolled up from a precisely cut strip of paper 12.7 mm wide and 152 mm long which is inserted on edge in a circular groove in a specially designed machine, several different makes of instrument being available. The crushing force is applied to the long edges of the sample until it is collapsed, the maximum load being recorded as the ring crush strength. Ten specimens are tested from both the machine and transverse directions of the sample.

A variety of tests are performed on corrugated board before it is made into the finished container, including bursting strength, edge compression, flat crush, pin adhesion and puncture.

The bursting strength of corrugated and solid fiberboard is described in TAPPI T 810 and is primarily an indication of the character of the materials used, giving no direct information regarding the ultimate performance of the finished container.

Edgewise compression strength of corrugated board gives information about the compressive strength, parallel to the flutes, of single-, double-, and triple-wall corrugated board. A specimen 31.75 × 50.8 mm (long edges perpendicular to the flute axis) is dipped into molten paraffin to a depth of 6.35 mm and then placed in a compression-testing instrument. The load required to failure is then determined, and this value can be related to the top-to-bottom compressive strength of the finished corrugated container. If the load is applied at an angle to the flute direction, the maximum load that the board can withstand decreases.

The flat crush test (performed only on single-wall board due to the difficulty with lateral movement of the liners in double- and triple-wall corrugated board) measures the resistance of the flutes in corrugated board to a crushing force applied perpendicular to the surface. From this test some indication of the durability of the finished corrugated board container is obtained. A circular specimen about 3225 mm^2 in size is placed between two parallel platens in a compression-testing instrument, and the force required to crush the sample is measured. Low flat crush values indicate poor flute formation or bad corrugating medium.

The pin adhesion test evaluates the strength of the adhesive bond and is probably the most commonly test performed on corrugated board. It measures the force required to separate the bond between the flutes and the linerboard, using pins (attached to a special fixture) which are inserted edgewise into a small rectangular sample of corrugated board along the open spaces in the flutes. The entire fixture is then placed in the jaws of a compression tester and a separating force applied.

Tests performed on the finished container include the puncture, compression resistance, impact resistance, stiffness, and drop tests. Of these, the compression test is the most common.

The compression test measures the ability of the container (usually empty but occasionally with inner packing or fully loaded) to resist external compressive forces and is used by many corrugated paperboard converters and box manufacturers as a routine quality control test. The container can be positioned in various ways to measure the resistance: top-to- bottom, end-to-end, diagonally opposite edges or any other desired direction. The sealed container is placed in the center of the bottom platen of the compression instrument and force applied at a constant rate; readings are taken until failure of the container occurs.

As Peleg has pointed out, specification of container strength by compression test rather than board bursting strength would be more valuable since the main concern of container users is the stacking strength of the container rather than the material it is made of. A main obstacle to the use of compression tests is the lack of performance standards specifying compression test tolerances for optimal performance of the various containers.

Chapter 3

The Kinds of Deteriorative Reactions

3.1 Introduction

Knowledge of the kinds of deteriorative reactions that influence food quality is the first step in developing food packaging that will minimize undesirable changes in sensory properties and maximize the development and maintenance of desirable properties.

Once the nature of the reaction is understood, a knowledge of the factors that control the rates of these reactions is necessary in order to fully control the changes occurring in foods during storage, *i.e.*, while packaged. The nature of the deteriorative reactions in foods is reviewed in this section, and the factors which control the rates of these reactions are discussed in the following section.

Let us try to understand few more details about enzymes

Enzymes are complex, globular proteins that have three important characteristics. They can act as catalysts, accelerating the rate of chemical reactions by factors of 10^{12}–10^{20} over that of uncatalyzed reactions. A second characteristic is their specificity, enabling the food processor to selectively modify individual food components and not affect others. As well, they can have a regulatory role, controlling the biochemical processes associated with various membranous components of a tissue, thus having a profound effect on regulating life processes and defining the quality of foods during postharvest or postmortem processing and storage.

During cheese-making (*e.g.*, rennin) and those produced by added and adventitious microorganisms, combine to produce the desired characteristic flavors and textures of the cheese. Enzymes supplied by microorganisms (adventitious or intentionally added) can have profound roles in altering flavors, colors and textures in food during intentional fermentation (desirable), and during food spoilage (undesirable).

Enzymes endogenous to plant or animal tissues or biological fluids can have undesirable or desirable consequences in foods. Examples involving endogenous enzymes include autolysis of fish postharvest, the senescence and spoilage of fruits and vegetables postharvest, glycolysis in postmortem, prerigor muscle and suberization (wound healing) of plant tissues postharvest.

Examples involving endogenous enzymes include oxidation of lipids by lipases and lipoxygenase (affecting the color, flavor and texture), oxidation of phenolic substances in plant tissues by phenolase (leading to browning), hydrolysis of phospholipids in fish by phospholipases (leading to changes in texture), sugar-starch conversions in plant tissues by amylases, invertase and invertase inhibitor, and postharvest demethylation of pectic substances in plant tissues (leading to softening of plant tissues during ripening, and firming of plant tissues during processing).

The importance of enzymes to the food processor is often determined by the conditions prevailing within and outside the food. Control of these conditions is necessary to control enzymic activity during food processing and storage. The major factors useful in controlling enzyme activity are temperature, water activity, pH, chemicals which can inhibit enzyme action, alteration of substrates, alteration of products, and preprocessing control.

Three of these factors are particularly relevant in a packaging context. The first is temperature: the ability of a package to maintain a low product temperature and thus retard enzyme action will often increase product shelf life. The second important factor is water activity, since the rate of enzyme activity is dependent on the amount of water available; low levels of water can severely restrict enzymic activities and even alter their pattern of activity. Finally, alteration of substrate (in particular, the ingress of oxygen into a package) is important in many oxygen-dependent reactions which are catalyzed by enzymes.

Table 3.1: Chemical Reactions that can Lead to Deterioration of Food Quality or Impairment of Safety

Nonenzymic browning

Lipid hydrolysis

Lipid oxidation

Protein denaturation

Protein cross-linking

Oligo- and polysaccharide hydrolysis

Protein hydrolysis

Polysaccharide synthesis

Degradation of specific natural pigments

Glycolytic changes

Many of the chemical reactions occurring in foods can lead to a deterioration in food quality (both nutritional and sensory) or the impairment of food safety. The more important classes of these reactions are discussed fully in standard textbooks on food chemistry. Worthy of note in the present context is the fact that such reaction classes can involve different reactants or substrates depending on the specific food and the particular conditions for processing or storage.

The rates of these chemical reactions are dependent on a variety of factors amenable to control by packaging including light, oxygen concentration, temperature and water activity. Therefore, the package can in certain circumstances play a major role in controlling these factors, and thus indirectly the rate of the deteriorative chemical reactions.

3.2 Major Changes in Food Quality

The two major chemical changes which occur during the processing and storage of foods and lead to a deterioration in sensory quality are lipid oxidation and non-enzymic browning. Chemical reactions are also responsible for changes in the color and flavor of foods during processing and storage.

Autoxidation is the reaction of molecular oxygen by a free radical mechanism with hydrocarbons and other compounds. The reaction of free radicals with oxygen is extremely rapid, and many mechanisms for initiation of free radical reactions have been described. Lipids which contain multiple double-bonded systems such as polyunsaturated fatty acids and phospholipids are particularly susceptible to autoxidation. The crucial role which autoxidation plays in the development of undesirable flavors and aromas in foods is well documented, and is a major cause of food deterioration.

The mechanism of lipid oxidation is characterized by four major steps initiation, propagation, branching and termination. Initiation takes place by loss of a hydrogen radical due to heat, light or trace metals. In propagation, the lipid-free radical reacts with oxygen to form peroxy free radicals, which in turn react with more lipid molecules to form hydro-peroxides. During the branching process, there is a geometrical increase in free radicals from decomposition of hydroperoxide. Termination involves the elimination of free radicals by addition of two free radicals or transfer of the radical to a compound to form a stable radical. The four processes are summarized in Table 3.2. RH refers to any unsaturated fatty acid in which the hydrogen is labile by reason of its position on a carbon adjacent to a double bond. Termination occurs through recombination of radicals to produce relatively unreactive products.

As well as being responsible for the development of off-flavors in foods, the products of lipid oxidation may also react with other food constituents such as proteins, resulting in extensive cross-linking of the protein chains through either protein-protein or protein-lipid crosslinks.

Table 3.2: Four Major Steps Characterizing Autoxidation of Lipids

Initiation:	$RH \rightarrow R\bullet + H\bullet$
Propagation:	$R\bullet + O_2 \rightarrow ROO\bullet$
	$ROO\bullet + RH \rightarrow ROOH + R\bullet$
Branching:	$ROOH \rightarrow RO\bullet + OH \rightarrow \bullet 2RH + ROH + H_2O$ (monomolecular decomposition)
	$2ROOH \rightarrow ROO\bullet + RO\bullet + H_2O$ (bimolecular decomposition)
Termination:	$ROO\bullet + ROO\bullet \rightarrow ROOR + O_2$
	$R\bullet + R\bullet \rightarrow R\text{–}R$
	$R\bullet + ROO\bullet \rightarrow ROOR$

Factors which influence the rate and course of oxidation of lipids are well known and include light, local oxygen concentration, high temperature, the presence of catalysts (generally transition metals such as iron and copper, but also heme pigments in muscle foods) and water activity. Control of these factors can significantly reduce the extent of lipid oxidation in foods.

Nonenzymic browning is one of the major deteriorative chemical reactions which occurs during storage of dried and concentrated foods. Although the reactions involved are complex and, as yet, not completely elucidated, the non-enzymic browning or Maillard reaction can be divided into three stages:

1. Early Maillard reactions which are chemically well-defined steps without browning;

2. Advanced Maillard reactions which lead to the formation of volatile or soluble substances; and

3. Final Maillard reactions leading to insoluble brown polymers.

The initial reaction involves a simple condensation between an aldehyde (usually a reducing sugar) and an amine (usually a protein or amino acid) to give a glycosylamine via a Schiff's base. The glycosylamine then undergoes an Amadori rearrangement to form an Amadori 1-amino-1-deoxy-2-ketose derivative.

The formation of Amadori compounds accounts for the observed loss of both reducing sugar and amine during the Maillard reaction. Although the early Maillard reactions forming Amadori compounds do not cause browning, they do reduce nutritive value.

There are five pathways in the advanced Maillard reactions. The first two pathways start from the 1–2 enol or 2–3 enol forms of the Amadori product and yield various flavor compounds. The third pathway is the Strecker degradation, which involves oxidative degradation of amino acids by the dicarbonyls produced in the first two pathways. The fourth pathway involves transamination of the Schiff base, and the final pathway starts with a second substitution of the amino-deoxy-ketose.

The final step of the advanced Maillard reaction is the formation of many heterocyclic compounds such as pyrazines and pyrroles, as well as brown melanoldin pigments. These pigments are formed by polymerization of the reactive compounds produced during the advanced Maillard reactions, such as unsaturated carbonyl compounds and furfural. The polymers are relatively inert and have a molecular weight greater than 1000.

Acceptability of color in a given food is influenced by many diverse factors, including cultural, geographical and sociological aspects of the population. However, regardless of these many factors, certain food groups are only acceptable if they fall within a certain color range. The color of many foods is due to the presence of natural pigments. The nature and major changes which these can undergo are briefly described; a more detailed discussion can be found elsewhere.

Chlorophylls

The name chlorophyll describes those green pigments involved in the photosynthesis of higher plants. In general terms, chlorophylls are magnesium-chelated tetrapyrrole structures with various substitutions (most commonly methyl) at different positions. The major change which chlorophylls can undergo is pheophytinization, the replacement of the central magnesium by hydrogen and the consequent formation of a dull olive-brown pheophytin. Because this reaction is accelerated by heat and is acid catalyzed, it is therefore unlikely to be influenced by the choice of packaging.

Almost any type of food processing or storage causes some deterioration of the chlorophyll pigments. Although pheophytinization is the major change, other reactions are possible. For example, dehydrated products such as green peas and beans packed in clear glass containers undergo photooxidation and loss of desirable color.

Heme Pigments

Meat is an important part of many diets, and the color of red meat is due to the presence of the heme pigment myoglobin. Myoglobin is a complex muscle protein contained within the cells of the tissues where it acts as a temporary storehouse for the oxygen brought by the hemoglobin in the blood.

3.3 Biological Changes in the Food Quality

The biological changes which occur in food are due to microorganisms.

Microorganisms can make both desirable and undesirable changes to the quality of foods, depending on whether or not they are introduced as an essential part of the food preservation process (*e.g.*, as inocula in food fermentations) or arise adventitiously and subsequently grow to produce food spoilage. In the latter case, they only reach readily observable proportions when they are present in the food in large numbers. Since the initial population or microbial load is usually small, observable levels are only reached after extensive multiplication of the microorganism(s) in the food.

The two major groups of microorganisms found in foods are bacteria and fungi, the latter consisting of yeasts and molds. Bacteria are generally the fastest growing, so that in conditions favorable to both, bacteria will usually outgrow fungi. The phases through which the two groups pass are broadly similar: a period of adjustment or adaptation (known as the lag phase) is followed by accelerating growth until a steady, rapid rate (known as the logarithmic phase since growth is exponential) is achieved. After a time the growth rate slows until growth and death are balanced and the population remains constant (known as the stationary phase). Eventually, death exceeds growth and the organisms enter the phase of decline.

Foods are frequently classified on the basis of their stability as non-perishable, semiperishable and perishable. An example of the first classification is sugar; provided it is kept dry, at ambient temperature and free from contamination, it should have a very long shelf life. However, few foods are truly nonperishable, and an important factor in determining their perishability is their packaging.

For example, hermetically sealed and heat processed (*e.g.*, canned) foods are generally regarded as nonperishable. However, they may become perishable under certain circumstances when an opportunity for recontamination is afforded following processing. Such an opportunity may arise if the can seams are faulty, or if there is excessive corrosion resulting in internal gas formation and eventual bursting of the can. Spoilage may also take place when the canned food is stored at unusually high temperatures: thermophilic spore-forming bacteria may multiply, causing undesirable changes such as flat sour spoilage.

Low moisture content foods such as flour, dried fruits and vegetables, and baked goods are classified as semiperishable. Frozen foods, though basically perishable, may be classified as semiperishable provided that they are properly stored at freezer temperatures.

The majority of foods (*e.g.*, flesh foods such as meat and fish; milk, eggs and most fruits and vegetables) are classified as perishable unless they have been processed in some way. Often, the only form of processing which such food receive is to be packaged and kept under controlled temperature conditions.

The species of microorganisms which cause the spoilage of particular foods are influenced by two factors: the nature of the foods and their surroundings. These factors are referred to as intrinsic and extrinsic parameters.

Table 3.3: Intrinsic and Extrinsic Parameters Influencing Microbial Growth in Foods

Intrinsic Factors	Extrinsic Factors
pH	storage temperature
a_w	relative humidity of
E_h	environment
nutrient content	Presence and concentration
antimicrobial constituents	of gases in the
biological structures	environment

Most microorganisms grow best at pH values around 7.0, while few grow below pH 4.0. Bacteria tend to be more fastidious in their relationships to pH than molds and yeasts, with the pathogenic bacteria being the most fastidious.

The minimum water activity (a_w) values reported for growth of some microorganisms in foods are presented in Table 3.4. Worthy of note is the fact that yeasts and molds grow over a wide a_w range than bacteria.

Table 3.4: Approximate Minimum a_w Values for the Growth of Microorganisms of Importance in Foods

Organism	Minimum a_w
Most spoilage bacteria	0.91
Most spoilage yeasts	0.88
Most spoilage molds	0.80
Halophilic bacteria	0.75
Xerophilic molds	0.65
Osmophilic yeasts	0.60

Certain relationships have been shown to exist between a_w, temperature and nutrition. Firstly, at any temperature, the ability of microorganism to grow is reduced as the a_w is lowered. Secondly, the range of a_w over which growth occurs is the greatest at the optimum temperature for growth; thirdly, the presence of nutrients increases the range of a_w over which the organisms can survive.

The oxidation-reduction potential (E_h) of a substrate may be defined as the ease with which the substrate loses or gains electrons. The more highly oxidized a substance, the more positive will be its electrical potential. Aerobic microorganisms require positive E_h values (oxidized) for growth while anaerobes require negative E_h values. Among the substances in foods that help to maintain reducing conditions are the sulfhydryl groups on proteins, and ascorbic acid and reducing sugars in fruits and vegetables. Deteriorative chemical reactions can alter the E_h value of foods during storage.

In order to grow and function normally, microorganisms require several nutrients: water, a source of energy, a source of nitrogen, vitamins and related growth factors, and minerals. The availability of water is related directly to the water activity of the food. The primary source of nitrogen utilized by microorganisms is amino acids. Growth factor and vitamin requirements tend to be specific to individual groups of microorganisms. Some foods contain certain naturally occurring substances which have been shown to have antimicrobial activities, thus preventing or retarding the growth of specific microorganisms in those foods.

The extrinsic parameters of foods are those properties of the storage environment that affect both the foods and their microorganisms. The growth rate of the microorganisms responsible for spoilage primarily depends on these extrinsic parameters: storage temperature, relative humidity and gas composition of the surrounding atmosphere.

The temperature of storage is particularly important, and several food preservation techniques (*e.g.*, chilling) rely on reducing the temperature of the food to extend its shelf life. Although there is a very wide range of temperatures over which the growth of microorganisms has been reported, specific microorganisms have relatively narrow temperature ranges over which growth is possible. Those which grow well below 20°C and have their optimum between 20° and 30°C are referred to as *psychrophiles* or *psychrotrophs*. Those that grow well between 2° and 45°C with optima between 30° and 40°C are referred to as *mesophiles,* and those that grow well at and above 45°C with optima between 55° and 65°C are referred to as *thermophiles.* Molds are able to grow over a wider range of temperature than bacteria, many being capable of growth at refrigerator temperature. Yeasts grow over the psychrophilic and mesophilic temperature ranges but generally not within the thermophilic range.

The relative humidity (RH) of environment is important and can influence the water activity of the food unless the package provides a barrier. Many flexible plastic packaging materials provide good moisture barriers but none are completely impermeable, thus limiting the shelf life of low a_w foods.

The presence and concentration of gases in the environment has a considerable influence on the growth of microorganisms. Increased concentrations of gases such as CO_2 are used to retard microbial growth and thus extend the shelf life of foods. As well, vacuum packaging (*i.e.*, removal of air (and thus oxygen) from a package prior to sealing) can also have a beneficial effect by preventing the growth of aerobic microorganisms. This type of packaging (known as modified atmosphere packaging or MAP) raises certain safety issues. Most food pathogens do not grow at refrigerator temperatures, and carbon dioxide is not highly effective at non-refrigeration temperatures. Therefore, most MA packaged food is usually held under refrigeration. Temperature abuse of the product (*i.e.*, holding at non-refrigerated temperatures) could allow the growth of organisms (including pathogens) which had been inhibited by carbon dioxide during storage at lower temperatures. For these reasons, it is difficult to evaluate MAP safety solely on the growth of certain pathogens at abusive temperatures.

3.4 The Common Insect Pests

The common insect pests of fresh food are flies (from the order Diptera), and cockroaches. They are attracted by food odors regardless of whether the food is fresh or beginning to decay. Any insect in food is a pest since not only does the food become contaminated with their bodies and excreta, but they are also capable of transmitting pathogens, including food poisoning organisms.

In contrast, the main insect species important as pests of stored foods are entirely from the orders *Lepidoptera* (moths) and *Coleoptera* (beetles). They regularly damage and destroy large quantities of stored foods around the world every year. The number of species involved is not large, there being no more than about twelve really important ones. They include weevils, various other beetles and the larvae of several moths. Most stored-product insects are cosmopolitan in that any given species is for the most part found worldwide in areas with similar climatic conditions. Warm humid environments promote insect growth, although most insects will not breed if the temperature exceeds about 35°C or falls below 10°C. Also many insects cannot reproduce satisfactorily unless the moisture content of their food is greater than about 11 per cent.

Moths and beetles are generally found in dry storage areas. They are able to survive on very small amounts of food, and thus can persist on food residues in improperly cleaned premises or equipment. Good ventilation, the use of cool storage areas and rotation of stock assists in keeping these pests at bay. Basic to effective control of insect pests is an understanding of their life cycles and feeding habits.

In common with many insects, moths pass through four stages during their development: the egg, the larva (caterpillar), the pupa (chrysalis) and the adult moth, The larval stage is the only one which consumes food; their presence can be recognized by a characteristic mixture of silken threads and frass (droppings) which they produce. Beetles have the same four life stages as moths, but they differ in that the adult beetle is a considerably harder-bodied insect than the moth, and may live and feed for months or even years. Cockroaches are larger, more robust insects which are highly mobile. Their young ones are like small versions of the adult and unlike beetles and moths, they do not inhabit packaged foods. Ants are a highly specialized group of insects which form nests, normally outside buildings. Worker ants may travel considerable distances and collect almost any type of food (especially sweet or high-protein foods) and take it back to the nest.

Table 3.5: Resistance of Various Materials to Insect Penetration

	Excellent	Good	Fair	Poor
Polycarbonate	x			
Poly(ethylene terephthalate)	x			
Cellulose acetate		x		
Polyamide		x		
Polyethylene (0.254 mm)		x		
Polypropylene (blaxially oriented)		x		
Poly(vinyl chloride) (unplasticized)		x		
Acrylonitrile			x	
Poly(tetrafluoroethylene)			x	
Polyethylene (0.123 mm)			x	
Regenerated cellulose				x
Corrugated paperboard				x
Ethylene vinyl acetate copolymer				x
Ionomer				x
Kraft paper				x
Paper/foil/polyethylene laminate pouch				x
Polyethylene (0.0254–0.100 mm)				x
Poly(vinyl chloride) (plasticized)				x
Vinyl chloride/vinylidene chloride copolymer				x

Mites which sometimes occur in stored foods are not insects but are closely related to spiders, having eight legs. They are minute in size, requiring a lens to see them, and are primarily pests of cereals and other foods with a moisture content of at least 12 per cent. Mites are so small that the presence of a few would pass unnoticed. They produce a sour odor in the food.

The main categories of foods subject to pest attack are cereal grains and products derived from cereal grains, other seeds used as food (especially legumes), dairy products such as cheese and milk powders, dried fruits, dried and smoked meats, and nuts. As well as their possible health significance, the presence of insects and insect excreta in packaged foods may render products unsalable, causing considerable economic loss, as well as reduction in nutritional quality, production of off-flavors, and acceleration of decay processes due to the creation of higher temperatures and moisture levels.

Early stages of infestations are often difficult to detect due to the small size of the insects and the fact that some species are only active at night. However, infestations can generally be recognized not only by the presence of the insects themselves but also by the products of their activities such as webbing, clumped-together food particles and holes in packaging materials. Each pest species tends to have a particular preference as to the kind of material attacked, so that the species that occurs in any particular situation depends very largely on the nature of the food involved.

Unlike microorganisms, some insects species (penetrators) have the ability to bore through one or more of the flexible packaging materials in use today and take up residence inside. Other species (invaders) usually do not enter packages unless there is an existing opening. However, such openings need not be very large. For example, the adult sawtoothed grain beetle can enter an opening less than 1 mm in diameter. Newly hatched larvae can enter much smaller openings; holes only 0.1 mm in diameter are sufficient to admit the larvae of some insects. Thus, package seals are critical in the design and use of packages that are expected to protect foods from insect infestations.

Unless plastic films are laminated with foil or paper, insects are able to penetrate most of them quite easily, the rate of penetration usually being directly related to film thickness. In general, thicker films are more resistant than thinner films, and oriented films tend to be more effective than cast films. The looseness of the film has also been reported to be an important factor, loose films being more easily penetrated than tightly fitted films. Regenerated cellulose, polyester and polyethylene films are inadequate barriers to insects if used singly; however, if laminated with biaxially oriented polypropylene, increased resistance can be obtained.

Generally, the penetration varies depending on the basic resin from which the film is made, on the combination of materials, on the package structure, and on the species and stage of insects involved. Absolute values are difficult to determine because resistance to penetration is influenced by factors such as package configuration and the presence or absence of folds, tucks and other harborage sites. Therefore, after appropriate packaging materials have been selected they must be evaluated *in situ* for insect resistance.

Aluminum foil (0.10 mm thick) has been shown to be penetrated by cadelle larvae and by *Trogodenna inclusum,* a dermestid. Lamination of foil to either paper or plastic films usually improves its resistance to insect penetration. More recently it has been demonstrated that vacuumized polyethylene bags of shelled peanuts were penetrated by the flour beetle *Tribolium castaneum,* whereas unvacuumized bags of the same material, and both vacuumized and unvacuumized bags made from poly(ethylene terephthalate) film were not. The author had no ready explanation for the differences in behavior of the various packs.

In an evaluation of films for their intrinsic capacity to exclude insects, it was noted that impregnation of these materials with pesticides conferred a greater degree of resistance. However, only pyrethrin synergized with piperonyl butoxide is registered in the USA for use as an insecticide for package treatment. Packaging applications that are currently permitted are large cotton bags, the outer ply of large multiwall paper bags, and two-ply regenerated cellulose/polyolefin sheets, in which case the insecticide combination is added to the adhesive.

Insects readily infest paper and can find ideal conditions for growth and and reproduction within this material. Synergized pyrethrins have been added to paper at levels of 16.4 mg m^{-2} in multiwall paper bags. Corrugated paper shipping containers, although not directly in contact with food, may greatly influence the potential for contamination of foods in individual packages as it provides ideal shelter for insects to hide away in.

**Table 3.6: Methods for Obtaining Insect Resistance
of Packaging Materials**

1.	Select a film and film thickness that are inherently resistant to insect penetration.
2.	Use a film adhesive containing an approved insecticide.
3.	Add an insecticide directly to the packaging material.
4.	Use shrink film overwraps to provide an additional barrier.
5.	Seal carton flaps completely.

The rodents rats and mice are among humanity's most cunning and capable enemies. They have highly developed senses of touch, smell and hearing, and can identify new or unfamiliar objects in their environment. Rats can wriggle through openings the size of a quarter; a mouse needs a hole only as large as a nickel to gain access. Rats and mice carry disease-producing organisms on their feet and/ or in their intestinal tracts and are known to harbor salmonellae of serotypes frequently associated with foodborne infections in humans. In addition to the public health consequences of rodent populations in close proximity to humans, these animals also compete intensively with humans for food.

Rats and mice gnaw to reach sources of food and drink and to keep their teeth short. Their incisor teeth are so strong that rats have been known to gnaw through lead pipes and unhardened concrete, as well as sacks, wood, and flexible packaging materials. Obviously proper sanitation in food processing and storage areas is the most effective weapon in the fight against rodents, since all packaging materials apart from metal and glass containers can be attacked by rats and mice.

3.5 The Influence of Water on Chemical Reactivity

Water may influence chemical reactivity in different ways. It may act as a reactant or as a solvent, where it may exert a dilution effect on the substrates, thus decreasing the reaction rate.

Water may also change the mobility of the reactants by affecting the viscosity of the food systems, and form hydrogen bonds or complexes with the reacting species. Thus, a very important practical aspect of a_w is to control undesirable chemical and enzymic reactions that reduce the shelf life of foods. It is a well-known generality that rates of changes in food properties can be minimized or accelerated over widely different values of a_w. Small changes in a_w can result in large changes in reaction rates.

The influence of a_w on lipid oxidation has been studied extensively, mainly with the use of model systems. The general effect of a_w on lipid oxidation is shown in Figure 3.1. At very low a_w levels, foods containing unsaturated fats and exposed to atmospheric oxygen are highly susceptible to the development of oxidative

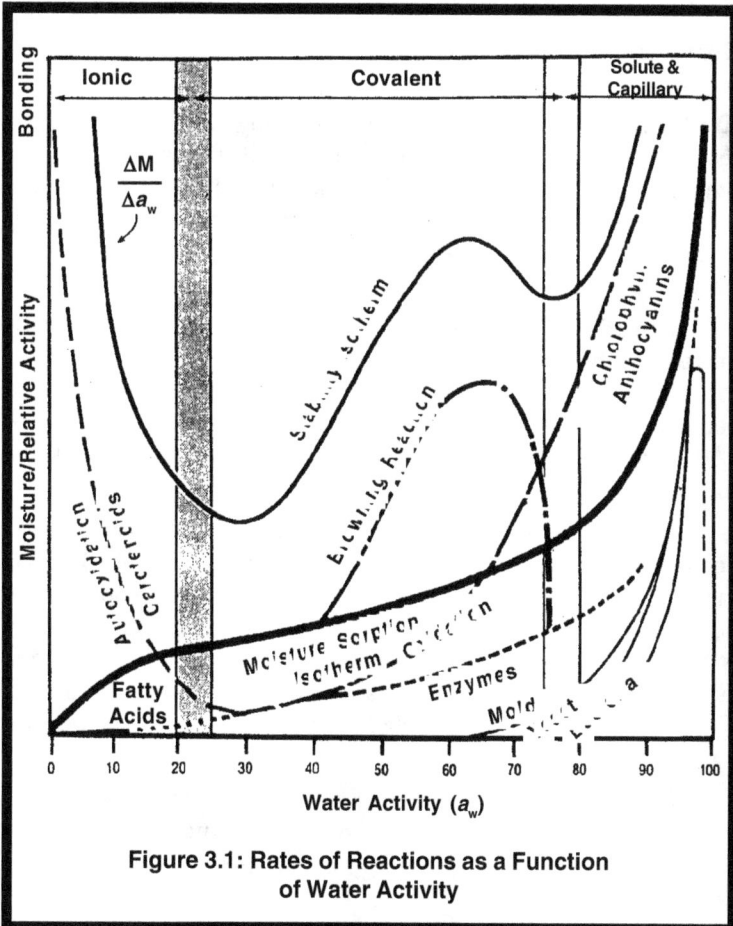

Figure 3.1: Rates of Reactions as a Function of Water Activity

rancidity. This high oxidative activity occurs at a_w levels below the monolayer level, and as a_w increases, both the rate and the extent of autoxidation increase until an a_w in the range of 0.3–0.5 is reached. Above this point, the rate of oxidation increases until a steady state is reached, normally at a_w levels in excess of 0.75. At a_w below the monolayer value, the oxidation rate decreases with increasing a_w. The rate reaches a minimum around the monolayer value and increases with further increase in a_w. Water may influence lipid oxidation by influencing the concentrations of initiating radicals present, the degree of contact and mobility of reactants, and the relative importance of radical transfer versus recombination reactions.

Browning

Water may accelerate browning by imparting mobility to the substrates, or it may decrease the browning rate by diluting the reactive species. In the low a_w range, the mobility factor predominates, whereas the dilution factor predominates in the high a_w range. As a consequence, browning rate generally increases with increasing a_w at low moisture content, reaches a maximum at a_w's of 0.4–0.8, and decreases with further increase in a_w.

Activation energies for the formation of Amadori compounds decrease with increasing a_w and level off at a_w of about 0.50. Therefore, the reaction rate is a_w highly temperature dependent at low a_w and becomes much less dependent on temperature at a_w greater than 0.50. This is the reason for the use of low air temperatures during the final stage of drying.

The effect of a_w on vitamin degradation in food systems has been reviewed. The rate of degradation of vitamins A, B_1, B_2 and C increased as a_w increased over the range 0.24–0.65. In contrast, the degradation of α-totocopherol in a fat-free model system increased with increasing a_w over the range 0.10–0.65. Generally, the rate of ascorbic acid degradation increases exponentially with increase in a_w. The photodegradation of riboflavin has been shown to increase with increasing a_w.

Near or below the monolayer a_w value, enzyme activities are generally minimized or cease. Above the monolayer value, enzyme activity increases with increasing a_w or increased substrate mobility,

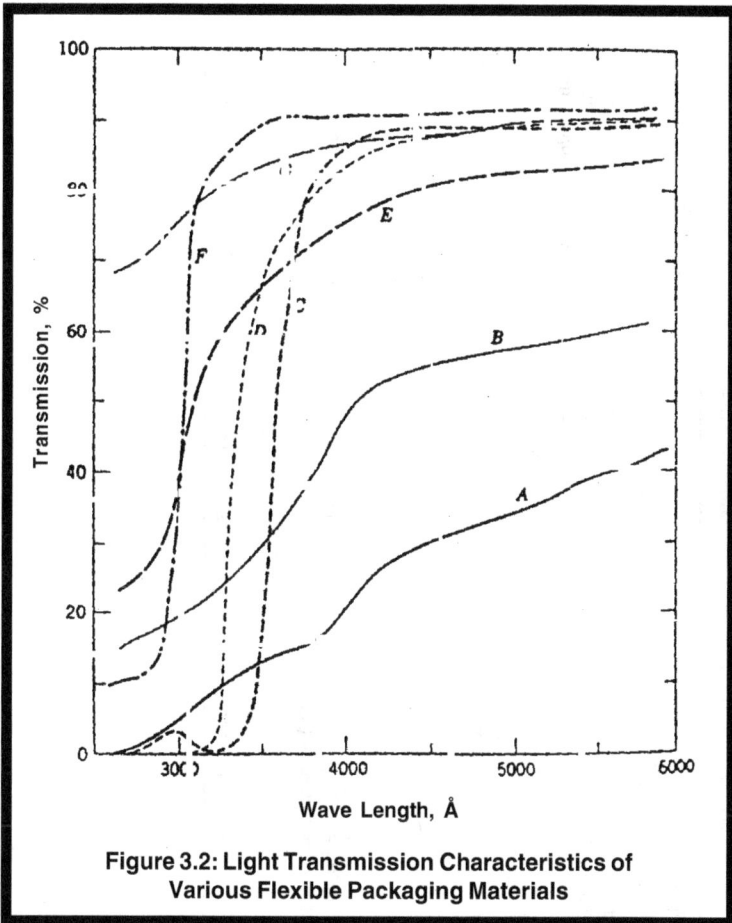

Figure 3.2: Light Transmission Characteristics of Various Flexible Packaging Materials

as illustrated in Figure 3.2. Substrates of high molecular weight such as protein and starch are less mobile than low molecular weight substrates such as glucose, and generally the latter have a lower a_w threshold for enzyme activity. At subfreezing temperatures, the reaction rate generally decreases with decreasing temperature and a_w, due partly to the lower temperature, partly to the increase in viscosity of the partially frozen system, and partly to enzyme denaturation.

Every microorganism has a limiting a_w value below which it will not grow, form spores or produce toxic metabolites. Table 3.7

Table 3.7: Water Activity and Growth of Microorganisms in Food

Range	Microorganisms Generally Inhibited by Lowest a_w in this Range	Foods Generally within this Range
1.00–0.95	Pseudomonas, Escherichia, Proteus, Shigella, Klebsiella, Bacillus, Clostrium pertingens, some Yeasts	Highly perishable (fresh) foods and canned fruits, vegetables, meat, fish and milk; cooked sausages and breads; foods containing up to approximately 40 per cent (w/w) sucrose or 7 per cent NaCl
0.95–0.91	Salmonella, Vibrio parahaemolyticus, C. botulinum, Serratia, Lactobacillus, Pediococcus, some molds, yeasts	Some cheeses (cheddar, Swiss, Muenster, provolone), cured meat (ham), some fruit juice concentrates; foods containing 55 (w/w) sucrose or 12 NaCl
0.91–0.87	Many yeasts (Candida, Torulopsis, Hansenula), Micrococcus	Fermented sausage (salami), sponge cakes, dry cheese, margarine; foods containing 65 per cent (w/w) sucrose (saturated) or 15 per cent NaCl
0.87–0.80	Most molds (mycotoxigenic penicillia), Staphylococcus aureus, most Saccharomyces (bailii) spp., Debaryomyces	Most fruit juice concentrates, sweetened condensed milk, chocolate, syrup, maple and fruit syrups; flour, rice, pulses containing 15–17 per cent moisture; fruit cake, country-style ham, fondants, high-ratio cakes
0.80–0.75	Most halophilic bacteria, mycotoxigenic aspergilli	Jam, marmalade, marzipan, glaced fruits, some marsh-mallows

Contd...

Table 3.7–Contd...

Range	Microorganisms Generally Inhibited by Lowest a_w in this Range	Foods Generally within this Range
0.75–0.65	Xerophilic molds (*Aspergillus chevalieri, A. Candidus, Wallemia sebid*), *Saccharomyces bisporus*	Rolled oats containing approximately 10 per cent moisture, grained nougats, fudge, marshmallows, jelly, molasses, raw cane sugar, some dried fruits, nuts
0.65–0.60	Osmophilic yeasts (*Saccharomyces rouxii*), few molds (*Aspergillus echinulatus, Monascus bisporus*)	Dried fruits containing 15–20 per cent moisture, some toffees and caramels; honey
0.50	No microbial proliferation	Pasta containing approximately 12 per cent moisture; spices containing approximately 10 per cent moisture
0.40	No microbial proliferation	Whole-egg powder containing approximately 5 per cent moisture
0.30	No microbial proliferation	Cookies, crackers, bread crusts, and so on containing 3–5 per cent moisture
0.20	No microbial proliferation	Whole milk powder containing 2–3 per cent moisture, dried vegetables and corn flakes containing approximately 5 per cent moisture, country-style cookies, crackers

lists the range of water activity permitting the growth of various common microorganisms, together with common foods categorized according to their a_w.

Water activity can influence each of the four main growth cycle phases by its effect on the germination time, the length of the lag phase, the growth rate phase, the size of the stationary population, and the subsequent death rate. Generally, reducing the a_w of a given food increases the lag period and decreases the growth rate during the logarithmic phase, the maximum of which becomes lower.

Sporulation may occur at or slightly below the minimum a_w for growth. In contrast, germination of spores of some microorganisms may occur at a_w values below that required for growth. The minimum a_w for growth of microorganisms is, without exception, less than or equal to the minimum a_w for toxin production. Optimal conditions of temperature, pH, oxygen tension and nutrient availability are necessary to permit sporulation, germination and toxin production at reduced a_w.

Whether a microorganism survives or dies in a low a_w environment is influenced by intrinsic factors that are also responsible for its growth at a higher a_w. These factors include water-binding properties, nutritive potential, pH, E_h and the presence of antimicrobial compounds. The influences exerted by these factors interact with a_w both singularly and in combination. Microbial growth and survival are not entirely ascribed to reduced a_w but also to the nature of the solute. However, the exact nature of the role that water plays in the mechanism of cell survival is not clearly understood.

Key extrinsic factors relative to a_w that influence microbial deterioration in foods include temperature, oxygen and chemical treatments. These factors can all combine in a complex way to either encourage or discourage microbial growth. Detailed discussion of these inter-relationships is beyond the scope of this book, and the interested reader is referred to standard textbooks on food microbiology.

Although the water vapor permeability of the packaging is a decisive factor in controlling changes in moisture content and thus water activity of packaged foods, the a_w of food in impermeable packaging can alter as a consequence of microbial growth. For example, experiments with slices of cake contaminated with pure strains of xerotolerant (*i.e.*, tolerant of dry conditions) molds stored in water-impermeable packaging showed two phases in the a_w evolution when molds grew. During the first stage, a_w decreased significantly, probably due to water uptake by spores and water binding in the young cytoplasm. In the second phase, the respiration of the growing mycelium produced metabolic water in excess and the a_w rose, progressively allowing the growth of less xerotolerant species which were present.

Chapter 4

Food Packaging Metals and their Corrosion

4.1 Introduction

The chemical structure of metals which gives them their valuable practical properties is also responsible for their main weakness and their susceptibility to corrosion.

Corrosion is the term used to describe the chemical reaction between a metal and its environment to form compounds; it is a universal process affecting all metals to a greater or lesser extent. Because the reaction takes place at the metal surface, the rate of attack can be reduced and controlled by modifying the conditions at the surface.

Metals are chemically reactive and can be readily oxidized by oxygen and other agents to form largely useless corrosion products. This vulnerability to oxidation accounts for the fact that with few exceptions (copper, silver and gold), metals do not occur naturally in the metallic state but are found combined with oxygen or sulfur in their ores. A considerable amount of energy is required to extract metals from their ores, and the reverse process (which releases energy) is strongly favored as the metal reverts back to its natural state. As a very broad generalization it can be said that the more difficult it has been to win the metal from its natural form, the greater will be its tendency to return to that form by corroding, but the rate of return will of course depend on the environment.

The reaction of metals in aqueous solutions or under moist conditions (known as wet corrosion) is electrochemical in nature, involving the transfer of electric charge across the boundary formed by the metal surface and its environment. An electrolyte is a medium which conducts electricity by movement of ions, the cations (*e.g.,* Fe^{2+}) and anions (*e.g.,* Cl^-) moving in opposite directions. When an electrode reaction takes places at a metal surface, the electron flow in the metal corresponds to an ion flow in the electrolyte.

When a metal corrodes, atoms of the metal are lost from the surface as cations, leaving behind the requisite number of electrons in the body of the metal. This dissolution of the metal is called an anodic reaction and takes place at a surface termed an anode; an anodic reaction involves the release of electrons or electrochemical oxidation. Thus, for the case of a metal M:

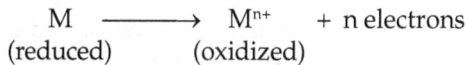

$$M \longrightarrow M^{n+} + n\ electrons$$
$$(reduced) \qquad (oxidized)$$

Simultaneously, reagents in the electrolyte solution react with the metal surface to remove electrons left behind by the departing metal ions. This removal of electrons is termed a cathodic reaction and takes place at a surface called a cathode. The cathodic reaction always involves consumption of electrons or electrochemical reduction.

Since practically all metals are covered with an oxide film, this must be removed before the metal can be exposed to an electrolyte. A metal covered with an oxide has different properties in a solution from a bare metal, but in studying electrochemical corrosion, it is the simplest to begin with the ideal case of a pure, bare metal electrode on which the only reaction occurring is metal dissolution.

4.2 Factors Affecting the Rate of Corrosion

The factors that affect the rate of corrosion can be studied under the following headings:

1. Polarization of the electrodes
2. Supply of oxygen
3. Temperature
4. Passivity

Polarization of the Electrodes

When a current flows, there is a change in the potential of an electrode; this is known as polarization. As the current begins to flow, the potential of the cathode becomes increasingly negative and the anode increasingly positive. Consequently the potential difference between the anode and the cathode decreases until a steady state is reached when corrosion proceeds at a constant rate. Thus, the corrosion current and therefore, the corrosion rate will be affected by anything which affects the polarization of the electrodes.

The potential at which the reaction takes place changes by an amount called the overpotential, η, which is defined as:

$$\eta = E_{corr} - E_i$$

where E_i is the polarized potential. The anodic overpotential η_a drives the metal dissolution process, and the cathodic overpotential η_c drives the cathodic deposition process.

Overpotential increases with current density, i, in accord with the Tafel equation:

$$\eta = \beta \log \frac{i}{i_o}$$

where β and i_o are constants for a given metal and environment; both depend on temperature. The dominant polarization term controlling the corrosion rate of many metals in deaerated water is the hydrogen overpotential at cathodic areas of the metal. The hydrogen overpotential for iron at 16°C in 1 N hydrochloric acid is 0.45 V, and for tin at 20°C in 1 N hydrochloric acid is 0.75 V.

Supply of Oxygen

The rate of which oxygen is supplied largely governs the rate of corrosion, since corrosion by oxygen reduction requires the presence of oxygen for the cathodic reaction to proceed. The rate of supply is proportional to the rate at which oxygen diffuses to the metal surface, and this depends on the concentration of dissolved oxygen in solution. This is further justification for the practice of attempting to remove all the oxygen from canned foods prior to seaming on the can end.

Temperature

The rate of corrosion generally increases with increase in temperature as more reactant molecules or ions are activated and are able to cross over the energy barrier. As well, increasing the temperature tends to increase the rate of diffusion of molecules or ions in a solution, although the solubility of oxygen in water decreases with increasing temperature.

Passivity

The dissolution process a metal was described as an oxidation process of the general form $M \rightarrow M^n + n$ electrons. From the Pourbaix diagrams, it is evident that if the metal can be oxidized to an oxide that is stable in the electrolyte, then the metal is rendered passive (*i.e.*, passivated). Passivation usually requires strong oxidizing conditions.

Thus, corrosion resistant metals and alloys can withstand an aggressive environment because of the presence of thin films of adherent oxides on their surfaces. The oxide layer will completely stop the anodic reaction which is the direct cause of corrosion, and if the film is insoluble in the electrolyte solution, it will form an insulation barrier which will reduce the rate of the cathodic reaction.

For example, iron is readily attacked by dilute nitric acid, but is inert in concentrated nitric acid because a thin, protective film is formed. As a result, iron behaves in concentrated nitric acid like a much more noble metal than it actually is. Iron can also be passivated by chromate solutions, as can tinplate, the latter being a very important step in the manufacture of tinplate. Passivation of tinplate can be achieved using an aqueous solution of chromic acid, although an electrolytic treatment in a sodium dichromate electrolyte has gained widespread favor. The resultant film is composed of chromium and chromium oxides and tin oxide, its properties varying depending on the quantity and form of these basic components.

4.3 Corrosion of Plain Tinplate Cans and Enameled Cans

The tinplate surface consists of a large area of tin and tiny areas of exposed tin-iron alloy ($FeSn_2$) and steel as a result of pores and scratches in the tin coating. Although the now obsolete hot dipped tinplate had a substantial tin-iron alloy layer, electrolytic tinplate has a much thinner layer which is electropositive to the base and also to the tin, thus acting as a chemically inert-barrier to attack on the steel base. The effect of this barrier is to prevent a significant increase in the steel cathode area. Thus, the density or degree of continuity of the alloy layer has a material effect on the rate of corrosion. The alloy-tin couple test gives a good indication of the continuity of the alloy layer in tinplate.

In the case of tinplate exposed to an aerated aqueous environment, tin is noble (*i.e.*, cathodic) to iron according to the electrochemical series. Therefore, all the anodic corrosion is concentrated on the minute areas of steel and the iron dissolves, *i.e.*, rusts. In extreme cases, perforation of the sheet may occur. This is the process which occurs on the external surface of tinplate containers.

However, inside a tinplate can, the tin may be either the anode or the cathode depending on the nature of the food. In a dilute aerated acid medium the iron is the anode and it dissolves, liberating hydrogen. In deaerated acidic food, iron is the anode initially, but later reversal of polarity occurs and the tin becomes the anode, thus protecting the steel; tin has been described in this situation as a sacrificial anode. This reversal occurs because certain constituents of foods can combine chemically with Sn^{2+} ions to form soluble tin complexes. As a consequence, the activity of Sn^{2+} ions with which the tin is in equilibrium is greatly lowered, and the tin becomes less noble (*i.e.*, more electropositive) than iron.

By using the Nernst equation, it is possible to calculate the ratio of Sn^{2+} to Fe^{2+} within the can for the reversal of polarity to occur. For the system 0.1 M citric acid at pH 3.8:

$$Fe^{2+} + Sn \longrightarrow Sn^{2+} + Fe$$

$$E = [0.14 - 0.44] - \frac{0.0592}{2} \log \frac{(Sn^{2+})}{(Fe^{2+})}$$

The cell reverses polarity when E = 0. Hence:

$$\log \frac{(Sn^{2+})}{(Fe^{2+})} = -0.30 \times \frac{2}{0.0592} = -10.30$$

and

$$Sn^{2+} = 5 \times 10^{-11} \, Fe^{2+}$$

This very low concentration of Sn^{2+} relative to Fe^{2+} can only occur through the formation of tin complexes and these are discussed later. Generally, complexing agents tend to increase the corrosion rates of many metals by reducing the metal ion activity, thereby shifting metal potentials markedly in the active direction.

The reversal has also been related to the high hydrogen overpotential of tin compared with that of iron, and to the low solubility constant of stannous hydroxide (5×10^{26}) which forms and precipitates on the cathode and reduces its size, compared with that of iron (1.6×10^{15}).

Now let us study about corrosion of enameled cans.

Food cans with enamel (lacquer) coatings are used to protect against excessive dissolution of tin, sulfide staining, local etching and change in color of pigmented products such as berry fruits. However, the use of enamels will not guarantee the prevention of corrosion, and in some cases may actually accelerate it. Therefore, careful consideration must be given before selecting an enamel system for a particular canned food.

The general pattern of corrosion in enameled cans is very different from that in plain cans, and is generally more complex. It depends not only on the quality of the base steel plate, the tin-iron alloy layer and the tin coating, but also on the passivation layers and the nature of the enamel coating. The only exposure of metal in an enameled can is at pores and scratches in the enamel coating and at cracks along the side seam. Some of these discontinuities in the enamel coating may coincide with pores in the tin coating, thus resulting in exposure of the steel. Even if defects in the enamel film expose only the tin coating, the availability of all the corrosion promoters in the can for attack on the limited areas of tin ensures that steel is soon exposed at them. Because these areas of exposed steel are almost unprotected either by the electromechanical action of a tin coating or by dissolved tin, corrosion may proceed at a rapid rate, resulting in hydrogen swelling or perforation of the can. Thus, it is easily possible for the use of an enamel to actually reduce the shelf life of a canned product.

The effectiveness of an enamel coating is related directly to its ability to act as an impermeable barrier to gases, liquids and ions, thereby preventing corrosive action on the protected surface. The transport of ions through the enamel is governed by the electrochemical characteristics of the film, in contrast to the transport of gases and liquids which involves dissolution in the enamel film and diffusion through it under a concentration gradient.

Because ions are electrically charged, their transfer through the enamel coating (actually the flow of an electric current) is complex and depends not only on their electric charge but also of the concentration of the electrolyte. If the transport rate of cations and anions through the coating differs, the coating itself may become charged. Thus, the protection offered by the enamel coating depends on its resistance to ion transfer which may take place even in the absence of pores, scratches or blisters.

The performance of enameled food cans is greatly affected by the thickness of the enamel coating. A thickness of 4 to 6 µm is sufficient for nonaggressive products such as apricots and beans, but aggressive products such as tomato paste require thicknesses of 8 to 12 µm, the heavier coatings having much lower porosities.

A possible electrochemical corrosion mechanism in enameled cans has been proposed and a schematic diagram is presented as Figure 4.1. In a plain can, the product would attack the tin layer, causing it to dissolve. The presence of an enamel coating protects the surface, and tin dissolution occurs only from both sides of a scratch under the enamel or through a pore, causing anodic undermining. Because only a very small area is in contact with the product in the can, dissolution is slow. With time the exposed area increases, and detachment of the enamel may be observed, resulting in the appearance of enlarged pores in the enamel coating. The tin-iron alloy layer may also be visible as a grayish color. Since (in theory) the alloy layer is more passive than tin and iron, it should dissolve only after all the tin and iron. However, because of its extreme thinness and its coupling effect with the tin, the potential of the can may be more noble than that of iron, which may also go into solution.

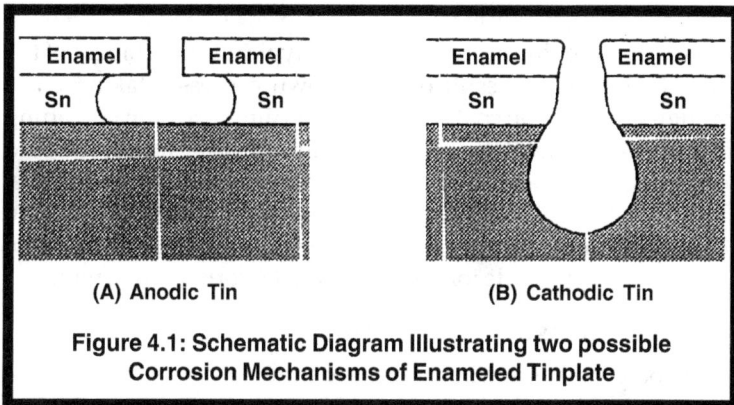

(A) Anodic Tin   (B) Cathodic Tin

Figure 4.1: Schematic Diagram Illustrating two possible Corrosion Mechanisms of Enameled Tinplate

In this situation, the enamel coating may serve as the cathode due to diffusion of protons through it.

Mannheim and Passy have used tomato concentrate as an example of the above scenario: while the overall rate of tin dissolution is reduced in the presence of an enamel coating, protection of the steel by the tin is also impaired due to reduction in the ratio of exposed tin to steel areas. Corrosion is concentrated in small areas and strong local currents may occur. The presence of corrosion accelerators such as nitrates aggravates the situation.

However, since no tin is dissolved, the alloy layer is not laid bare and thus, has no influence on the corrosion process. Failure occurs eventually due to pinhole formation (also known as pitting corrosion) but there is no undermining of the enamel, corrosion usually starting at a point of discontinuity such as a scratch or pore in the coating. Aggressive products such as beets in acetic acid and berries are responsible for this type of corrosion. The cans do not appear to be corroded, and it is only on closer inspection (usually with a, hands lens) that spots of corrosion (often with deep penetration into the steel) are visible.

Failure of enameled cans is often due to a reduction in the bond between the enamel and the metal surface, resulting in eventual lifting of the enamel coating. Thus, good adhesion is required to prevent anodic reactions, to counteract forces developed under the coating due to physical or chemical factors, and to ensure an aesthetic appearance.

Loss of adhesion may commence prior to corrosion due to stresses induced during fabrication, but typically detachment of the enamel coating is a result of breakdown processes taking place through or under the coating, or to underfilm corrosion spreading from exposed metal. The quality of the surface of the tinplate prior to enameling has a major effect on adhesion and enamel performance. The presence of salt residues can cause failure due to their ability to attract water and therefore, establish a conducting film beneath the coating, and to provide ions for carrying the corrosion current.

4.4 The Corrosion Accelerators

Food products and beverages are extremely complex chemical systems covering a wide range of pH and buffering properties, as well as a variable content of corrosion inhibitors or accelerators, Factors which influence the corrosiveness of food products and beverages can be divided into two groups hartwell: intensity and type of corrosive attack inherent in the food itself, and corrosiveness due to the processing and storage conditions, All these factors are interrelated and may combine in a synergistic manner to accelerate corrosion.

The most important corrosion accelerators in foods include oxygen, anthocyanins, nitrates, sulfur compounds and trimethylamines.

From a corrosiveness point of view, it is convenient to divide foods into five classes: those that are highly corrosive such as apple and grape juices, berries, cherries, prunes, pickles and sauerkraut; those that are moderately corrosive such as apples, peaches, pears, citrus fruits and tomato juice; those that are mildly corrosive such as peas, corn, meat and fish; and strong detinners such as green beans, spinach, asparagus and tomato products; beverages are conveniently considered as a fifth class.

While the above classification offers a good, broad guide, it is important to note that different lots or varieties of the same food can exhibit as much variation in their corrosiveness as may exist between different types of foods, Thus, for example, the same variety of fruit from different growing regions may vary several fold in corrosiveness. According to Hartwell, finding a twofold or greater variation between different packs of the same food in identical containers is the usual experience of investigators in this field.

The various factors that can influence the corrosiveness of food products and beverages will now be considered in more detail.

Acidity

No direct proportionality exists between the acidity of a product and the degree of corrosion of tinplate. In other words, two products of the same acidity will not necessarily be equally corrosive. It also appears that pure solutions of organic acids are less corrosive than the fruit juices containing them, suggesting that fruit juices contain unidentified depolarizers which enhance the corrosive action of organic acids. It has been well established that the tendency of an acid to form a complex with dissolved tin has an important bearing on the relative polarity of tin and steel, and hence the degree of corrosion.

In a study comparing six organic acids in a model system, acetic and lactic acids were the most corrosive based on subjective severity of can pitting; the type of organic acid was more important than its concentration, and the addition of nitrate increased pitting severity in all acid solutions.

pH

As was the case with acidity, no direct proportionality exists between pH and the degree of corrosion of tinplate; this is not surprising when it is considered that the reaction product when a metal is dissolved is not always an ionic species but often a solid oxide or hydroxide. Pourbaix diagrams indicate the relationship between pH and the equilibrium potential, and show regions of immunity, corrosion and passivity. The pH of the system also determines the relative cathodic protection given to steel. In some cases, tin is cathodic to steel over a certain pH range (in the case of acetic acid the range is pH 2.0 to 4.5), while in others it offers protection up to pH 4; above that level it may accelerate corrosion.

Sulfur Compounds

Sulfur and sulfur compounds may be introduced into the can in a number of ways: as spray residues from agricultural chemicals; as residues from sulfur-containing preservatives, and as components in sulfur-containing compounds such as proteins in meat, fish and certain vegetables: the proteins are degraded during heat processing, releasing free sulfide or hydrosulfide ions and evolving hydrogen sulfide gas into the headspace.

Trace amounts of sulfur compounds from agricultural chemicals (for example, derivatives of thio- and dithiocarbamic acid fungicides) can lead to accelerated corrosion and failure of plain tinplate cans containing acid foods such as apricots and peaches. As well, pitting corrosion can occur, and this has been attributed to inactivation of the tin coating by a protective film of sulfide having a more cathodic potential. As a consequence, there is a significant reduction in the tin dissolution rate, and no electrochemical protection of the steel by the tin.

Sulfur dioxide may be directly reduced on the tin surface to sulfide or even to sulfur, with tin passing into solution and the development of unpleasant odors and flavors. Residual SO_2 accelerates corrosion through its action as a depolarizer, inducing a negative charge in the double layer: this repels the electrons from the electrode, thus shifting the potential in the positive direction. Trace amounts of SO_2 as low as 1 mg kg^{-1} are sufficient to accelerate corrosion; such corrosion problems may be overcome by the use of enameled cans.

**Table 4.1: Some Corrosion Promoting Agents
and their Mode of Reaction**

Corrosion Accelerator	Reduction Product	Equivalent in Weight
Proton (H⁺)	H_2	1 mL H_2 = 5.3 mg Sn^{2+}
Oxygen (O_2)	H_2O	1 mL O_2 = 10.6 mg Sn^{2+}
Sulfur dioxide (SO_2)	H_2S	1 mL SO_2 = 5.5 mg Sn^{2+}
Sulfur (S)	H_2S	1 mL S = 3.7 mg Sn^{2+}
Nitrate (NO_3)	NH_3	1 mg NO_3 = 7.65 mg Sn^{2+}
Trimethylamine oxide (TMAO)	TMA	1 mg TMAO = 1.57 mg Sn^{2+}

There are two types of sulfide staining: one involves iron sulfide and is sometimes called sulfide black, and the other involves tin sulfides. Neither type constitutes a health hazard nor leads to failure of the can, but both types may cause adverse reactions from the consumer.

Iron sulfide stains are characteristically black and usually occur at isolated points on the can (mainly in the headspace region) during or immediately after heat processing. Iron sulfide is not formed at pH values below about 6 so that it is uncommon to find it in the portion of the can in contact with the contents. However, the pH of condensed volatile matter in the headspace may be above 6. The problem may be overcome by using enameled cans or plain cans with enameled ends.

Tin sulfide staining is usually widespread throughout the can and is blue-black or sometimes brown. Two stages are believed to be involved: an oxidation of the tin, followed by deposition of an insoluble tin sulfide precipitate on the surface. It occurs during or soon after heat processing and shows little or no increase in intensity during storage. It may be reduced by the use of CDC plate or prevented by using sulfur-resistant enamels into which quantities of zinc oxide or carbonate are added before being applied to the plate surface. These react with sulfur-bearing gases to form almost invisible white zinc sulfide.

Nitrates

Nitrates are found in fruits and vegetables grown in heavily fertilized soils, and may also occur in water supplies as a result of pollution by fertilizers. Vegetables such as green beans, spinach, turnips, lettuce, beets and radishes have been shown to contain on occasion several thousand mg kg-l of nitrates. Nitrates are very efficient cathode depolarizers since they are capable of being reduced all the way to ammonia. They have been responsible for serious economic and toxicological problems in some canned foods, notably tomato products. Although nitrates and nitrites are also present as intentional additives in processed meats, they present no problem; this is because meat products are above the critical pH range for detinning to occur via the nitrate-tin reduction system.

According to Britton, the reduction of 60 mg of nitrate could account for the dissolution of 470 mg of tin; a tin coating of 11.2 gsm may be dissolved completely within 6 months by tomatoes containing 100 mg kg^{-1} of nitrate.

Nitrates act as electron acceptors, replacing the hydrogen evolution reaction with the electron-nitrate two-step reduction reaction and shifting the reaction towards increased tin dissolution. The reduction reactions involved are believed to be as follows:

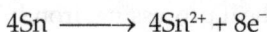

$$4Sn \longrightarrow 4Sn^{2+} + 8e^-$$

$$NO_3^- + 2e^- + 2H^+ \longrightarrow NO_2^- + H_2O$$

$$NO_2^- + 6e^- + 8H^+ \longrightarrow NH_4^+ + 2H_2O$$

The first reduction reaction equation is probably rate determining. The equations indicate that the rate of detinning depends on the nitrate concentration and pH. Nitrate does not immediately affect the corrosion rate, but begins to act after tin and iron ions have passed into solution. Although the ammonium ion is the major conversion product at pH 5 and less, above this pH other products are formed, including nitrous oxide (N_2O), nitric oxide (NO) and hydroxylamine. Oxygen present in the can at the time of processing triggers the nitrate-detinning reaction by increasing the initial rate of formation of Sn^{2+}.

Overcoming the problem by restricting the use of nitrate fertilizers has proved difficult, and instead efforts have been directed towards finding cultivars which do not accumulate high concentrations of nitrate. It is also possible to avoid the use of waters with a high nitrate content for canning operations; a suggested maximum is 5 mg L^{-1} nitrate. While raising the pH of the product above 6.0 will effectively inhibit nitrate detinning and can be used with some vegetables, such a procedure cannot be used with acid foods such as tomato or citrus products where the acidity determines the organoleptic properties and the heat treatment required. The best solution to the problem at present is the use of enameled cans, although the development of corrosion inhibitors (the search for such substances is continuing) would offer an alternative solution.

Phosphates

Phosphates are naturally present in meat and are often intentionally added as polyphosphates to processed meat products such as hams to reduce the loss of water during processing. The presence of phosphates leads to increased discoloration due to iron phosphate and sulfide formation. One countermeasure used against the effect of polyphosphate in ham cans has been the introduction of a small area of aluminum which acts as a sacrificial anode, protecting the tin surface. This is not always wholly successful, and parchment liners may be used in addition or as an alternative.

Plant Pigments

The anthocyanins and related pigments are among the most important potential corrosion accelerators (cathodic depolarizers) since they are easily reduced. Anthocyanin pigments can also act as anodic depolarizers through their ability to form complexes with cations, particularly those of iron and tin salts. Analysis of samples of canned fruit after a period of storage usually shows a greater amount of tin in the drained fruit than in the syrup, indicating that at least part of the tin is combined in an insoluble form with some constituent within the fruit.

The nature of the anthocyanin pigment is also important. For example, raspberries contain cyanidin glucosides which have *ortho*-dihydroxy groups in their structures; it is these groups which are involved in the formation of blue-tinted complexes with metals such as tin. The major pigment of strawberries is pelargonidin-3-glucoside which does not possess the necessary *ortho*-dihydroxy groups for complex formation. Therefore, strawberries do not show the same shift to a blue color in the presence of tin salts as do raspberries.

Combination of metal ions with tannins has been observed in other fruits. For example, discoloration in canned cranberry has been attributed to the formation of a complex between tannins present in the fruit and tin salts. Darkening in canned Maraschino cherries has been observed to be more severe when fruit of a high tannin content is used, a high tannin content being found in unripe fruit and fruit that has been stored in wooden barrels.

It should be noted that not all reactions between plant pigments and metal ions will produce undesirable colors, although this is usually the case with anthocyanin pigments.

Synthetic Colorings

The canned products which most commonly contain synthetic colorings are soft drinks, which consist basically of sugar-based syrups and carbonated water containing flavors, acidulants and colors. The behavior of soft drinks depends to a considerable extent on the presence of azo dyes (*e.g.*, amaranth) and the amount of residual oxygen in the filled can. Both of these components are capable of acting as corrosion accelerators and are potentially active corrosive agents. Tin dissolution may adversely affect the color of some products, and iron dissolution may lead to perforation and flavor defects. Thus fully enameled cans are essential, it being important to obtain near- perfect coverage by the enamel.

Soft drinks have been categorized as follows:

1. Products containing azo dyes or other oxidants which may promote dissolution of tin in the absence of oxygen.

2. Products based on citric acid where the tin is anodic and has a high dissolution rate.

3. Products based on phosphoric acid with fruit flavors where tin is strongly cathodic to iron and there is rapid corrosion of exposed steel because of the depolarizing action of the dyes.

4. Cola beverages containing phosphoric acid where the tin-iron couple action is insignificant; the tin serves as an inert barrier, resulting in relatively low iron pick-up.

5. Products based on citric acid without active dyes where tin is anodic to iron and reaction rates are generally slower. Protection of the exposed steel can be obtained by the deliberate addition of stannous chloride which acts as an effective corrosion inhibitor, but such addition is banned in many countries.

Copper

Foods containing dissolved copper will deposit it when put in a metallic container, and in acid products, this can lead to accelerated corrosion, either by tending to strip the tin or by producing local attack on the steel. The basic reason why copper accelerates the corrosion of steel is because copper catalyzes the reduction of oxygen.

In the past, copper all too frequently entered products by the solution of copper oxide from copper-bearing metal food contact equipment. However, with the replacement of much of this equipment with stainless steel equipment, the problem of copper in canned foods has been dramatically reduced. Another possible source of copper is from certain fungicides, and occasionally some cannery water supplies may have unacceptably high copper levels.

4.5 Effect of Processing and Storage

Oxygen

Oxygen may be dissolved in food products: it is naturally present in food tissues such as fruit and vegetable cells, and is inevitably entrained or absorbed by particulate products prior to their being filled into cans. Removal of as much of this oxygen as possible is an essential part of good cannery practice and a variety of methods are used: hot-filling, vacuum filling, exhausting, closure under vacuum, steam flow closure and vacuum syruping. As well, positive control of the headspace volume is essential.

A larger headspace is likely to contain a higher residual oxygen concentration than a smaller one. However, the larger the headspace, the more room there is for accumulation of hydrogen resulting from corrosion and thus, the greater the time required to form a hydrogen swell. In practice, the cannery technologist has little room to maneuver the headspace if the declared weight on the can label is to be met.

Despite the above procedures, there will always be some oxygen present in the headspace of newly filled cans. The rate of oxygen consumption at this stage is quite rapid but decreases with time, the rate being a function of initial concentration, headspace volume, can vacuum, nature of the product and type of container.

Oxygen acts as a depolarizer, accelerating corrosion by reacting with the hydrogen formed in the can through a cathodic reaction as shown in earlier equation. Calculations suggest that if the residual pressure in the headspace of a filled can of a common size is 0.5 atmospheres, then the oxygen present would be capable of removing about 0.25 gsm of tin from the coating, producing 15 mg kg^{-1} of dissolved tin in the can contents. Although this is a relatively small amount in comparison with the total coating weight, the action of the oxygen may be localized on a narrow zone of the gas/liquid interface in the headspace. This may result in the production of an unsightly etched line and even exposure of an area of steel much larger than that exposed at pores in the initial coating.

Thermal Processing

Little is known about the effect of heat sterilization process on corrosion rates, except that the quantity of metal dissolved during the process is very small. This is hardly surprising given the comparatively short processing times (typically within the range of 30 to 120 minutes) relative to the total shelf life of the canned product (typically from 1 to 2 years).

Storage Temperature

The rate of a chemical reaction increases as the temperature is raised, and for many reactions, the rate doubles for each 10°C increase in temperature, *i.e.*, the temperature quotient Q_{10} equals 2. Thus to minimize undesirable reactions such as nonenzymic browning in canned foods, it is preferable that storage temperatures be kept as low as practicable.

The situation with respect to the rate of corrosion of tinplate cans is less clear; data on the relationship between temperature and the corrosion current measured between tin and iron for some test solutions and food products showed a wide variation, and in all cases studied, Q_{10} was less than 2. For 3 per cent acetic acid, a negative Q_{10} was observed, since it is a tin dissolver at 20°C but an iron dissolver between 50 and 60°C. Similar reversal of the corrosion effect has been observed in fish products, but as a critical review commented "it is wrong to conclude from this that increasing the temperature improves storage stability, since corrosion is only one of the processes affecting shelf life."

Chapter 5
Packaging of Foods in Metal Containers

5.1 Introduction

The commercial packaging of foods in metal containers began in the early 19th century, following on from the discovery in the 1790s by the French confectioner Nicolas Appert of a method of conserving all kinds of food substances in containers a method to which the term canning is now applied indiscriminately, whether the container is made from, aluminium, glass or plastics.

Four metals have become commonly used materials for the packaging of foods: steel, aluminum, tin and chromium. Tin and steel, and chromium and steel, are used as composite materials in the form of tinplate and electrolytic chrome-coated steel (ECCS), the latter being somewhat unhelpfully referred to for many years as tin-free steel (TFS). Aluminum is used in the form of purified alloys containing small and carefully controlled amounts of magnesium and manganese. Two other metals are used during the soldering or welding of three-piece tinplate and ECCS containers: lead and copper. However, since they are not used for the fabrication of containers in their own right, they will not be discussed in detail in this chapter. The health aspects of these different metals, together with can coatings, are discussed elsewhere.

After almost two centuries of history, there is still controversy as to who introduced the tin can as a package. The 15th edition of the Encyclopaedia Britannica (1987) states in the section on food processing that "Peter Durand, an Englishman, conceived and patented the idea of using tin cans instead of bottles.....", in conflict with the British Dictionary of National biography's (1882–1900) assertion that Bryan Donkin "devised the method of preserving meat and vegetables in air-tight cases."

The first commercial manufacture of tinplate commenced in England in 1699 and in France in 1720 where it was used for a variety of purposes including household utensils such as plates. Some time about the middle of the eighteenth century, the Dutch navy began to use foods preserved by packing them in fat in tinned iron canisters. After cooking and while still hot, the material to be preserved was placed into the canister, covered with hot fat and the lid immediately soldered on, a forerunner of what today could be described as a form of aseptic canning! Records show that from 1772 to 1777 the Dutch Government supplied their navy (which had been sent out to Suriname (formerly Dutch Guiana in South America) to quell a revolt) with roast beef packed in this way.

Before the end of the eighteenth century the Dutch had also established a small industry to preserve salmon in a similar manner. Freshly caught salmon were cleaned, cooked in boiling brine, smoked over a wood fire for two days and then placed in a tinplated iron box. The spaces were filled up with hot salted butter or olive oil and a lid soldered onto the box. A famous London firm of snuff merchants supplied 13 tins of Dutch salmon to one of their clients in 1797. Thus, a canning industry of sorts had been established in Holland independently of Appert's work.

Durand's patent granted in 1810 speaks of "an invention communicated to him by a certain foreigner residing abroad" and covered the use of "glass, pottery, tin or other metals or fit materials." It is open to doubt whether the patent contributed anything other than the "tin" among a list of metals and materials for preserving vessels. Appert deliberately avoided tinplate in his early work because of the poor quality of the French product, according to the 4th (1831) and 5th (1858) editions of his book. However, the quality of tinplate in England was good and it was freely available.

The Durand patent was acquired by the Englishman John Gamble who was involved with Bryan Donkin in developing the Fourdrinier papermaking machine. Donkin had become interested in the tinning of iron as early as 1808 and in 1811, in association with John Hall, founder of the famous Dartford Iron Works, developed a process based on that of Appert. It is not clear how this occurred. The same year, using the name Durand (it is thought to avoid the legal problem of explaining the assignment of the patent) Donkin applied to the British admiralty for a test of his product. Although the kind of container used is not known, the first substantial orders were placed in 1814 with the London firm of Donkin, Hall and Gamble for meat preserved in tinplate canisters. By the 1820s canned foods were a recognized article of commerce in Britain and France.

As the review by Bishop demonstrates, flat assertions as to the origin of inventions are generally dangerous. To complete the historical record, William Underwood left London and arrived in New Orleans in 1817; he traveled up to Boston where he started a business preserving food in glass jars by Appert's method. In 1819 Thomas Kennsett, also from England, started a similar business in preserved foods in New York in partnership with his father-in-law Ezra Daggett. The first offering of preserved provisions in tin cans in America is assumed to be the announcement by Daggett and Kensett in the New York Evening Post of July 18, 1822, although it was not until 1825 that they took out a patent in which "vessels of tin" were mentioned.

The American Civil War afforded the opportunity for canning to become a great industry, and by the end of the war in 1865 the canners had increased their output sixfold. For many years the cans were made slowly and laboriously by hand. Both ends were soldered to the can with a hole about 25 mm diameter left in the top. After the can was filled through this hole, a metal disk was soldered into place. The mechanical roll crimping (commonly known as double seaming) of the can ends into a body with a soldered sideseam was patented in 1896 by Max Ams of New York, making it possible to develop high-speed equipment for the making, filling and closing of these cans.

Today, materials like tinplate and aluminum have become universally adopted for the manufacture of containers and closures for foods and beverages, due to largely several important qualities of these metals. These include their mechanical strength and resistance to working, low toxicity, superior barrier properties to gases, moisture and light, their ability to withstand wide extremes of temperature and provide ideal surfaces for decoration and lacquering.

5.2 The Use of Aluminum

Aluminum is the earth's most abundant metallic constituent, comprising 8.8 per cent of the earth's crust; only the nonmetals oxygen and silicon are more abundant. Alumina or aluminum oxide (Al_2O_3) is the only oxide formed by aluminum and is found in nature as the minerals corundum (Al_2O_3); diaspore ($Al_2O_3.H_2O$); gibbsite ($Al_2O_3.3H_2O$); and most commonly bauxite, an impure form of gibbsite.

Hans Christian Oersted, a Danish chemist, first isolated aluminum in 1825 using a chemical process involving potassium amalgam. Between 1827 and 1845 Friedrich Wohler, a German chemist, improved Oersted's process by using metallic potassium. In 1854 Henri Sainte-Claire Deville in France obtained the metal by reducing aluminum chloride with sodium. Aided by the financial backing of Napoleon III, Deville established a large-scale experimental plant and displayed pure aluminum at the Paris Exposition of 1885. In 1886 Charles Martin Hall in the United States and Paul Héroult in France independently and almost simultaneously discovered that alumina would dissolve in fused cryolite (Na_3AlF_6) and could then be decomposed electrolytically to a crude molten metal. A low-cost technique (the Hall-Héroult process) is still the only method used for the commercial production of aluminum, although new methods are under study.

Due to the chemical stability of its oxides, the energy requirements for smelting are extremely high. This has led to the production of aluminum in areas where cheap electrical power is available, but current trends suggest that the days of cheap power may be coming to an end. If aluminum smelters are required to pay the true market price for power, then the use of aluminum as a packaging material is almost certainly likely to decline quite sharply.

Today a typical aluminum process would be as follows: alumina is dissolved in cryolite in carbon-lined steel boxes called pots. A carbon electrode or anode is lowered into the solution and an electric current of 50–150 MA is passed through the mixture to the carbon-cathode lining of the pot. The current reduces the alumina into aluminum and oxygen, the latter combining with the anode's carbon to form carbon dioxide, while the aluminum (denser than cryolite) settles to the bottom of the pot.

Most commercial uses of aluminum require special properties that the pure metal cannot provide. Therefore, alloying agents are added to impart strength, improve formability characteristics and influence corrosion characteristics. A wide range of aluminum alloys are available commercially for packaging applications, depending on the container design and fabrication method being used.

Table 5.1: Chemical Composition of Some Commonly Used Aluminum Alloys

Alloy Type	Typical Usage	Si	Fe	Cu	Mn	Mg	Cr	Zn	Ti
1050	Foils and flexible tubes	0.25	0.4	0.05	0.05	–	–	0.03	–
1100		1.0	0.20	0.05	–	–	0.10	–	–
3003		0.6	0.7	0.70	1.5	–	–	0.10	–
3004	Beverage closures and D&I can bodies	0.30	0.7	0.25	1.5	1.3	–	0.25	–
5050		0.40	0.7	0.20	0.1	1.8	0.10	0.25	–
5182	Easy open beverages	0.20	0.35	0.15	0.5	5.0	0.10	0.25	0.10
8079		0.30	1.3	0.05	–	–	–	0.10	–

Commercially pure aluminum (types 1100 and 1050) is used for the manufacture of foil and extruded containers since it is the least susceptible to work hardening.

The general effect of several alloying elements on the corrosion behavior of aluminum has been reported as follows:

> 1. Copper reduces the corrosion resistance of aluminum more than any other alloying element and leads to a higher rate of general corrosion;

2. Manganese slightly increases corrosion resistance;

3. Magnesium has a beneficial influence and Al-Mg alloys have good corrosion resistance;

4. Zinc has only a small influence on corrosion resistance in most environments, tending to reduce the resistance of alloys to acid media and increase their resistance to alkalis;

5. Silicon slightly decreases corrosion resistance, depending on its form and location in the alloy microstructure;

6. Chromium increases corrosion resistance in the usual amounts added to alloys;

7. Iron reduces corrosion resistance and is probably the most common cause of pitting in aluminum alloys; a high iron content increases the bursting strength but reduces the corrosion resistance.

8. Titanium has little influence on corrosion resistance of aluminum alloys.

Compared with tinplate and ECCS, aluminum is a lighter, weaker but more ductile material that cannot be soldered.

5.3 The Profile of Can End

The can end is of complex design developed for optimum deformation behavior, the latter being dependent on plate thickness, the precise contour of the expansion rings and the countersink depth. It is important that the ends are able to deform under internal and external pressure without becoming permanently distorted. In effect they must act like diaphragms, expanding during thermal processing and returning to a concave profile when vacuum develops inside the can on cooling.

The ends are stamped on power presses from tinplate sheet, generally of high temper grade, which has usually been previously lacquered. After stamping the ends fall through the press into the curler to form the outside curl and diameter.

A lining or sealing compound is then applied into the seaming panel; the sealant used is based on natural or synthetic rubber and is dispersed in water or solvent. Its constituents are subject to stringent food regulations. The purpose of the sealant is to assist the formation

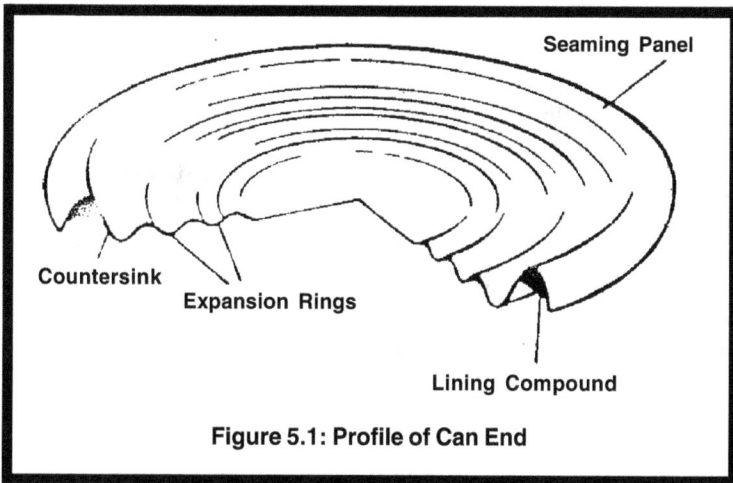

Figure 5.1: Profile of Can End

of an hermetic seal by providing a gasket between adjacent layers of metal.

Several types of easy-opening devices such as the key opening scored strip found in solid meat or shallow fish cans have been available for many years.

However, an increased demand for convenience features has seen the development of easy-open ends of two broad types: those which provide a pouring aperture for dispensing liquid products, and those which give a near full aperture opening for removing more solid products. The first easy-open end for canned beverages was developed in 1963.

Most designs incorporate a ring-pull end consisting of a scored portion in the end panel and a levering tab (formed separately) which is riveted onto a bubble-like structure fabricated during pressing. Modern designs (especially for beverage cans) have attempted to overcome litter problems by incorporating nondetachable tabs. Most but not all of the aperture circumference is scored to leave sufficient unscored portion to function as a hinge when the tab is pressed in.

Because of the greater ease of fabrication, integrated rivet ends have usually been made from aluminum but this presents problems when such ends are used with steel cans. For example, the corrosion of carbonated soft drinks in such cans is accelerated because of the bimetallic container, although the use of the same container for beer is beneficial because of the sacrificial nature of the aluminum and the more tolerable aluminum-beer reaction products than the iron counterpart. These problems have led to the development of steel easy-opening units, but they are much more difficult to fabricate.

Close control of scoring conditions is vital to ensure adequate resistance to bursting without requiring an unduly high tearing load to open. Particular attention must be paid to metal exposure resulting from internal enamel (lacquer) fracture at the score, and post-lacquer spraying is sometimes used to ovecome potential problems.

5.4 Double Seaming and Welded Side Seams

Welding has been used in the manufacture of steel pails for many years, but because of inconsistencies in weld integrity, the process was considered unacceptable for the production of high quality, leak-free food cans. Two processes were developed in about 1960 for welding side seams instead of soldering: the wire weld system developed by Soudronic A.G. of Switzerland, and the Conoweld roll-welding system developed by Continental Can Company in the United States. Both have since been improved and gamed widespread commercial acceptance.

Welding as an alternative to soldering spread widely in the 1980s due largely to the imposition by more and more countries of lower limits of lead in canned foods. Since the 1970s most countries have insisted that only pure tin solder be used on cans intended for baby foods; this added significantly to the cost of such cans and is one of the reasons that glass containers for baby food are so popular. However, in addition to the health aspects, welding offers savings in material, since the overlap needed to produce a weld uses less metal than an interlocked soldered seam. As well, the side seam is stronger, it is easier to seam on the ends, and a greater surface area is available for external decorating.

The Conoweld technique was developed initially for the production of beverage cans made from ECCS; it has since been expanded to tinplate aerosol containers and some food cans. To ensure satisfactory welds, thorough removal of the chrome/chrome oxide coating to a width of about 2 mm along each of the blank edges which will form the lap seam is required just prior to welding. A square wave form of alternating current is used to give solid-state welds which are forged together by a wheel.

The Soudronic Wima and Superwima (Wima is an abbreviation for wire mash) processes were developed in Switzerland for the high speed welding of tinplate and ECCS containers using a sine wave alternating current power supply and a continuous copper wire electrode to produce a weld with an extremely low metal overlap (less than 0.4 mm in the case of the Superwima seam). High electrical resistance causes the interface temperature to rise rapidly to at least

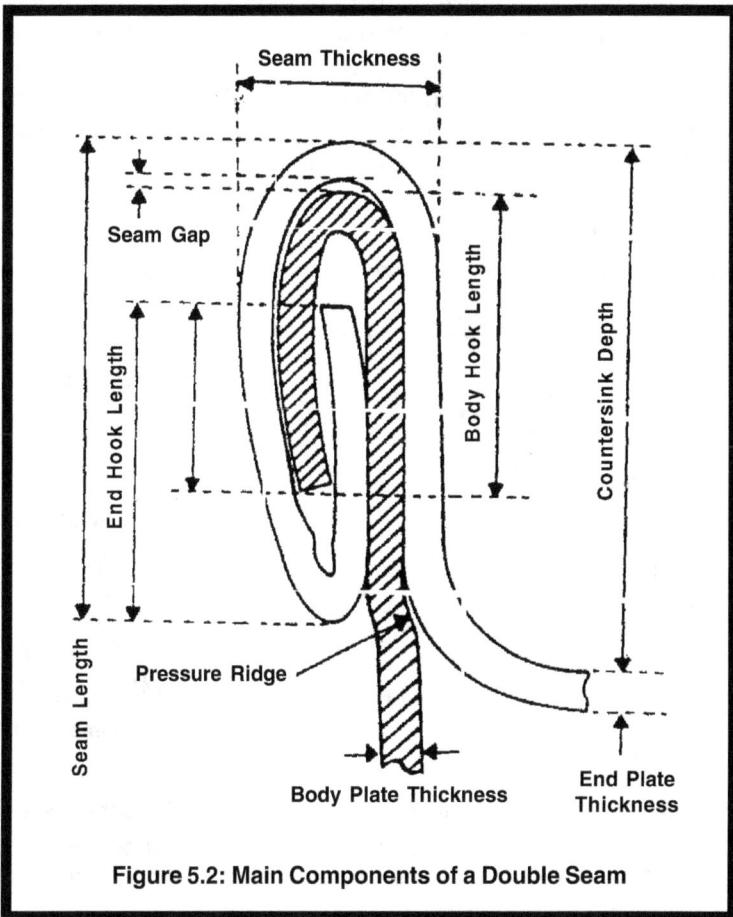

Figure 5.2: Main Components of a Double Seam

900°C and melt, resulting in solid phase bonding at all locations along the seam. The tensile strength of a good weld is equal to that of the base plate.

Although the welded seam is free from the danger of lead pick-up, the weld has to be effectively coated to prevent traces of iron being picked up by some types of beverages and acidic foods; side striping of the internal surface of the seam is required in these situations.

A system of chemical bonding of side seams has been developed by the American Can Company for ECCS can bodies. Known as the Miraseam process, it utilizes a thermoplastic polyamide adhesive which is applied to one edge of the pre-heated body blank before it is rolled into a cylinder, giving complete protection of the raw edges of the blank. A strong bonded lap seam is produced that is able to withstand the high in-can pressures generated by beers and carbonated soft drinks during can warming or pasteurisation. This method can only be used with ECCS since the melting point of tin is close to the fusion temperature of the plastic.

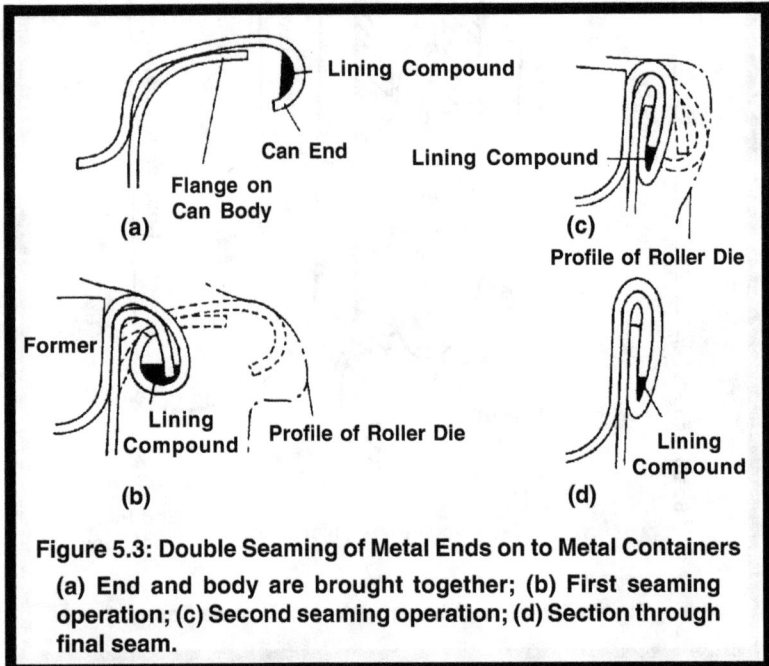

Figure 5.3: Double Seaming of Metal Ends on to Metal Containers
(a) End and body are brought together; (b) First seaming operation; (c) Second seaming operation; (d) Section through final seam.

The bodies are next transferred to a flanger for the final metal-forming operation: necking and flanging for beverage cans, and beading and flanging for food cans. The can rim is flanged outwards to enable ends to be seamed on. The top of beverage cans is necked to reduce the overall diameter across the seamed end to below that of the can body wall, yielding savings in the cost of metal through the use of smaller ends, and allowing more effective packing methods to be adopted. Simultaneous creation of the neck and flange using a spin process is now increasingly used. Double-, triple- and quadruple-necking is now quite common, the latter reducing the end diameter by over 8 mm (from 68.26 mm to 60.33 mm).

For food products where the can may be subjected to external pressure during retorting or remain under high internal vacuum during storage, the surface of the body may be beaded for strength. There are many bead designs and arrangements, all of which are attempts to meet certain performance criteria. Basically, beading produces shorter can segments that are more resistant to paneling, but such beads reduce the axial load resistance by acting as failure rings.

The end is then mechanically joined to the cylinder by a double seaming operation. This involves mechanically interlocking the two flanges or hooks of the body cylinder and end. It is carried out in two stages. In the first operation, the end curl is gradually rolled inwards radially so that its flange is well tucked up underneath the body hook, the final contour being governed by the shape of the seaming roll. In the second operation, the seam is tightened (closed up) by a shallower seaming roll. The final quality of the double seam is defined by its length, thickness and the extent of the overlap of the end hook with the body hook. Rigid standards are laid down for an acceptable degree of overlap and seam tightness.

Finally the cans are tested for leakage under pressure in large wheel-type testers; leaking cans are automatically rejected.

5.5 Aerosols and Aerosol Products

The word aerosol is used in colloid chemistry to describe a suspension in air or gas of small particles with a radius of less than 50 μm.

However, in the present context, an aerosol is defined as a self-contained sprayable product in which the expelling force is supplied by liquefied or compressed gases so that the product is self-dispensing.

The present aerosol industry is generally considered to have received its stimulus from the development of the aerosol insecticides used in the 2nd World War, although a few aerosols had been marketed prior to this time. Considerable information has been published about aerosols and patents granted over the period 1862 to 1942, and the reader is referred to standard texts for a detailed account of their development.

The history of aerosol food products can be traced back to 1925 when a method for fluffing and dispensing ice cream under pressure was patented by Ashley. However, long before that date, a patent was granted to Weyde and Matthews covering the use of nitrous oxide in artificially aerated drinks, and in 1882 the use of compressed gases for charging of milk products was patented. It was not until the mid-1950s that a range of food aerosol products appeared, including a drink mix, barbecue sauce, chocolate and table syrups, a vitamin syrup, a meat tenderizer and a salad dressing. Unfortunately, several of these products were disappointing and development work on new aerosol products dwindled almost to a stop.

In terms of the total worldwide aerosol business, the food aerosol market is very small. However, it is an area that does have significant growth potential. The major applications by volume at the present time appear to be whipped cream, cheese spreads and pan release sprays (*i.e.*, edible cooking lubricants).

Most aerosol products contain three major components: propellants, solvents and active ingredients. Aerosols can be divided into two broad classes (homogeneous or heterogeneous) depending on the manner in which these components are combined in the product. In homogeneous products, all the components are mutually soluble and two phases are present; a liquid phase in equilibrium with a vapor phase. Such products do not have to be shaken before use. Heterogeneous products are where the components are not mutually soluble and several types exist. Suspension systems consist of a solid phase and liquid phase in equilibrium with a vapor phase, and emulsion systems contain two liquid phases in equilibrium with a vapor phase. Virtually all food aerosols are heterogeneous, emulsion systems.

There are two main types of food aerosol systems on the market today: those where the propellant gas is mixed directly with the food system, and those where the propellant system is kept separate from the food product by means of an inert plastic bag or a moving piston device.

Propellant

Propellants provide the pressure that forces the product out of the container when the valve is opened. They also influence the form in which the product is discharged: foam, stream or spray. Variations in the type and concentration of a propellant can change a coarse, wet spray into a fine, dry spray, or a wet, sloppy foam into a dry, coherent foam.

There are three broad classes of propellants: hydrocarbons, fluorocarbons and compressed gases. For use in food aerosols, a propellant should be inert, non-toxic, non-flammable, odorless, tasteless and cheap. Several propellants are available for use in foods, including nitrous oxide (N_2O), carbon dioxide, nitrogen, and two fluorocarbons: Freon C-318 (octafluorocyclobutane) and Freon 115 (chloropentafluoroethane); these are the only two fluorocarbons that have been approved by the FDA for food use.

Freon 115 is very stable and has a high vapor pressure; it is used as a compressed gas or as a liquefied gas in combination with C-318 which acts as a vapor pressure depressant. Generally mixtures containing about 75/25 wt per cent of nitrous oxide-fluorocarbon have been found most satisfactory, greater foam stability resulting as higher percentages of fluorocarbon are included. However, actual commercial use of fluorocarbons in food aerosols is negligible; this is partly due to their much higher cost compared to the compressed gases, and partly to the role of chlorofluorocarbons in the destruction of the ozone layer. The pressure of the compressed gases changes relatively little with temperature in comparison to the liquefied gases.

For food aerosols, only the compressed gases are employed in any quantity, with nitrous oxide, either by itself or in combination with carbon dioxide, being the most common, the sweet taste of nitrous oxide offsetting the acidic taste of the carbon dioxide. Nitrogen is seldom used, for although it is inexpensive, stable and low in toxicity, it is almost completely insoluble in the liquid phase of aerosols, thus presenting a number of problems. Because it is practically insoluble in liquids, it concentrates in the vapor phase. As the product is discharged, the volume of the vapor phase increases with a consequent decrease in the pressure exerted by the nitrogen and a noticeable change in spray characteristics. Also, because nearly all of the nitrogen is in the vapor phase, the initial volume of this phase has to be considerably larger than is required for conventional liquefied gas propellants in order to have sufficient nitrogen to discharge the product completely. The same disadvantages apply to a lesser extent to carbon dioxide and nitrous oxide. Of course, an advantage of nitrogen is that because it is insoluble in the liquid phase, it does not alter the characteristics of the product. It is considered a good propellant for such products as pastes and syrups as it produces a solid stream.

Since the other compressed gases are soluble to only a limited extent in the liquid phase of aerosols, coarse sprays are produced and mechanical breakup actuators are necessary. The spray characteristics also change noticeably as the pressure drops during use, partly because the gas dissolved in the liquid phase is not released fast enough to maintain equilibrium between the liquid and vapor phases.

Misuse by the consumer of compressed gas aerosols by discharging the container when the end of the dip tube is in the vapor phase has been a major problem. If the valve is actuated when the container is in the inverted (and on occasions horizontal) position, the vapor phase is discharged instead of the liquid phase. This results in loss of the propellant, since most of the propellant is concentrated in the vapor phase, and the product cannot be obtained from the container. This problem is not as great with nitrous oxide and carbon dioxide as it is with nitrogen, because their increased solubility in the liquid phase provides a reservoir of propellant.

A modification of this system can produce solid-stream dispensing; a soluble compressed gas is used and the aerosol (without a dip tube) is used in the inverted position. When the valve is depressed, the material is emitted as a stream. This system was developed to be used with pressurized beverage products.

When selecting a propellant, the nature of the dispensed product must be considered since this is influenced by the type of propellants. The more soluble propellants (nitrous oxide and carbon dioxide) tend to form a product that foams, since when the product is dispensed, the dissolved gas expands and escapes into the atmosphere, resulting in dispersion of the liquid into fine particles. Nitrogen delivers the product in a nonaerated form.

The most common material used for aerosol containers is tinplate, but ECCS, aluminum, glass, stainless steel and plastics are also used to a lesser extent. Tinplate three-piece aerosol containers were traditionally soldered at the side seam but were one of the first groups of soldered metal containers to benefit from the conversion to welding. For food use, protective coatings as described previously for metal containers must be used.

For many food products, use of the typical aerosol systems may be unsuitable because of the viscosity of the product, incompatibility of the product and the propellant, or the desired dispensing characteristics of the finished product. In these situations, barrier containers have been designed. They have a specialized construction designed to keep the product completely separated from the propellant at all times from the initial filling to complete discharge. Two different types of containers are used for this purpose: one is a bag-in-can container and the other uses a plastic piston.

The main advantages of barrier packages are:

1. Products can be packaged without alteration of their composition or reformulation;
2. Very viscous products can be packaged;
3. The ratio of product to propellant is high (a consumer advantage);
4. Sterility is easier to achieve with barrier containers than with conventional aerosol systems;
5. The package can be discharged in any position.

The disadvantages of barrier packages is the higher cost, and they are not considered to be competitive with conventional aerosols.

Although all aerosol propellants are gases at room temperature and pressure, only the compressed gases are loaded into aerosol containers as gases. All other propellants are liquefied prior to loading, either by cooling them below their boiling point or increasing their pressure at room temperature. In the cold fill process, the cold active ingredient and cold propellant are both loaded into the container before the valve is added and crimped on. In the pressure fill process, the active ingredient is added to the container at room temperature and the valve crimped on; the propellant is then loaded through the valve under pressure.

When the compressed gas is added through the valve, the active ingredients are loaded in the container first, air is removed either by evacuation or purging with nitrogen, and the valve is added and crimped on. If carbon dioxide and nitrous oxide are employed as propellants, it is necessary to add a shaking step during the loading of the gas in order to establish equilibrium as rapidly as possible between the gas dissolved in the liquid phase and that in the vapor phase. Since nitrogen is almost insoluble in the active ingredients, it is forced through the valve without the need for shaking.

5.6 Coatings for Protection and Decorations

The primary function of interior can coatings is to prevent interaction between the can and its contents, although some enamels have special properties which allow products such as meat loaf to be easily removed from the cans, while others are used merely to improve the appearance of the pack. Exterior can coatings may be used to provide protection against the environment (*e.g.*, when the cans will be marketed in particularly humid or salt-laden climates), or as decoration to give product identity as well as protection.

There are several essential requirements needed in an interior coating: it has to act as an inert barrier, separating the container from its contents and not imparting any flavor to the contents; it must resist physical deformation during fabrication of the container and still provide the required chemical resistance; and the enamel must be flexible, spread evenly, completely cover the substrate and adhere to the metal surface. Adhesion failure may occur during, or as a result of, mechanical deformation during heat processing or undermining by corrosion.

For most containers the enamel is applied to the metal in the flat before fabrication, typical film masses being in the range 3–9 gsm (4–1.2 micron). However, because of the considerable amount of metal deformation with substantial disruption of the surface which takes place in the D&I operation, such containers must be coated internally after fabrication. Control of the dry film weight is essential if the enamel is to function correctly. A dry film which is too thin may not cover the surface completely, while an overthick film leads to brittleness and impaired protection, as well as being uneconomical. As the enamel film has imperfections and is damaged during container manufacture, it does not give complete protection to the can. Where it is essential to minimize product-container interactions, *e.g.,* for canned beer and soft drinks where metal pick-up can affect flavor and clarity, the cans are given a post-fabrication repair lacquering.

Many types of internal enamel are available for food containers including oleoresinous, vinyl, acrylic, phenolic and epoxy-phenolic. The original can lacquers were based on oleoresinous products which includes all those coating materials which are made by fusing natural gums and resins and blending them with drying oils such as linseed or tung. Although oleoresinous coatings are still used today, their open micellar structure means that they are prone to corrosion/staining problems with sulfur bearing products unless they are pigmented with zinc oxide. It was for this reason that a move has been made to synthetic phenolic resins dissolved in a blend of solvents.

Phenolic resins produced by the action of formaldehyde on phenol or other substituted phenols, have limited flexibility and high bake requirements, but are still used on three piece cans where flexibility is not required.

Vinyl coatings are based on copolymers of vinyl chloride and vinyl acetate of low molecular weight dissolved in strong ketonic solvents. The long carbon-carbon chains make them thermoplastic and they can be blended with alkyd, expoxy and phenolic resins to enhance their performance. Their flexibility allows them to be used for caps and closures as well as drawn cans. Their main disadvantage is their high sensitivity to heat and retorting processes, restricting their application to cans which are hot filled, and to beer and beverage products.

Epoxy phenolic coatings are made either by straight blending of a solid epoxy resin with a phenolic resin or are the products of the precondensation of a mixture of two resins in appropriate solvents. A three dimensional structure is formed during curing and baking which combines the good adhesion properties of the epoxy resin with the high chemical resistance properties of the phenolic resin. The balanced properties of epoxy phenolic coatings have made them almost universal in their application on food cans with the exception of deep multistage DRD cans.

Vinyl organosol coatings incorporate a high molecular weight PVC organosol dispersion resin which is thermoplastic in nature and extremely flexible. Soluble thermosetting resins (including epoxy, phenolic and polyesters) are added in order to enhance the film's product resistance and adhesion. Plasticizers are also added to aid film formation. These coatings are the most common on DRD cans in the USA and are typically white or buff colored due to the addition of titanium dioxide.

The types of resins most commonly used for internal coating of containers, their properties and general applications are summarized in Table 5.2. Generally the lacquers may be divided into three broad classes:

GP (General Purpose)

These are based on epoxy phenolic resins and are often used for the more acidic products. General purpose enamels are applied as either a single coat (they are then designated GP1), or as two coats (designated GP2). GP2 systems are used for products such as acidified beetroot and colored berry fruits which are especially corrosives. GP enamels also have some sulfur-resisting properties. GP enamels may be pigmented or have a clear appearance.

SR (Sulfur Resistant)

These enamels are used to prevent staining of tinplate surfaces by sulfur compounds released from foods such as meat, fish and vegetables which have sulfur-containing amino acids that breakdown during heat processing and storage to release sulfides. These react with tin to form black tin sulfide, or accumulate in the headspace and give out an unpleasant odor. To overcome this problem, two approaches have been used. Enamels are pigmented with zinc oxide which reacts with the sulfur compounds to form white zinc oxide (these are known as the sulfur-absorbing enamels), or the enamels are pigmented with aluminium powder or white pigment to obscure any tin sulfide which might form. These are known as sulfur-resisting enamels.

Special Enamels

Enamels having additives such as waxes to assist the release of

Table 5.2: Main Types of Internal Can Enamel (Lacquer)

General Type of Resin and Components Blended to Produce it	Flexibility	Sulfide Stain Resistance	Typical Uses	Comments
Oleo-resinous (drying oil and synthetic resins	Good	Poor	Acid fruits	Good general purpose natural range at relatively low cost
Sulfur-resistant oleo-resinous (added zinc oxide)	Good	Good	Vegetables, soups, especially can or as topcoat over epoxy-phenolic	Not for use with acid products
Phenolic (phenol or substituted phenol with formaldehyde)	Moderate	Very good	Meat, fish, soups, vegetables	Good at relatively low cost but film thickness restricted by flexibility
Epoxy-phenolic (epoxy resins with phenolic resins)	Good	Poor	Meat, fish, soups, vegetables, beer and beverages (top coat)	Wide range of properties may be obtained by modifications
Epoxy-phenolic with zinc oxide (zinc oxide added)	Good	Good	Vegetables, soups (especially can ends	Not for use with acid products; possible color change with green vegetables
Aluminized epoxy-phenolic (metallic aluminium powder added)	Good	Very good	Meat products	Clean but rather dull appearance

Contd...

Table 5.2–Contd...

General Type of Resin and Components Blended to Produce it	Flexibility	Sulfide Stain Resistance	Typical Uses	Comments
Vinyl solution (vinyl chloride-vinyl acetate copolymers)	Excellent	Not applicable	Spray on can bodies, roller coating on ends, as topcoat for beer and beverages	Free from flavor taints; sensitive to soldering heat and not usually suitable for direct application to tinplate
Vinyl organosol or plastisol (high MW vinyl resins suspended in a solvent)	Good	Not applicable	Beer and beverage topcoat on ends, bottle closures, drawn cans	As for vinyl solutions but giving a thicker, tougher layer
Acrylic (acrylic resin, usually pigmented white)	Very good in some ranges	Very good when pigmented	Vegetables, soups, prepared foods containing sulfide stainers	Attractive clean appearance of opened cans
Polybutadiene (hydrocarbon resins)	Moderate–poor	Very good if zinc	Beer and beverage first coat, vegetables and soups if with ZnO	Costs and hence popularity depends on country

the product from the can, or enamels pigmented with aluminum powder or other materials are also used. The latter were described above as sulfur-resisting enamels, but they are also used in premium quality packs where sulfur staining is not a problem, simply to improve the appearance of the inside can surface.

Two basic methods are used for the application of protective coatings to metal containers: roller coating and spraying, the former being the most widespread. Dipping was used in the early days but is now obsolete. Roller coating is used if physical contact is possible; thus it finds use in the coating of material in sheet and coil form, and the external coating of cylindrical can bodies. Spraying techniques are used if physical contact is impossible or difficult; thus it finds use mainly to coat the inside surface of can bodies, including two-piece D&I and sometimes DRD cans.

Because the coating is generally applied wet (*i.e.*, the resin is suspended in a carrier such as an organic or aqueous solvent for ease of application), it must be dried after application by solvent removal, oxidation and/or heat polymerization. This process (known as baking or curing) is usually carried out in a forced convection oven using hot air at up to 210°C for times of up to 15 minutes. However, recent developments include resin formulations which need lower temperatures and shorter curing times, often through the use of ultra-violet radiation to accelerate polymerization. Such UV-cured resins are virtually solvent-free; they contain photosensitive molecules which absorb the UV radiation and release free radicals which polymerize or cross-link the liquid resin to form a solid coating. Because the substrate is not heated, it can be handled immediately for further operations, and substantial savings (up to 50 per cent) in energy and space are possible.

Another recent development is that of power coating, where the resin is applied "dry" in the form of a fine powder, usually under the direction of an electrostatic field. It is used mainly where heavy coatings are required, such as in the protection of welded side seams where the bare metal that exists in the weld area (most of the tin is removed during the welding process) must be covered. Curing is usually by infrared radiation or high frequency induction heating, since hot air could disturb the uncured coating. Powdered coatings contain little or no volatile effluents and require low energy consumption for application and cure.

Electrophoretic deposition, a process originally developed for protecting automobile bodies, has been developed for coating two-piece cans. A resin film is deposited electrically from an aqueous suspension, giving a far more even distribution than that obtained by spraying. As well, its throwing power enables it to coat regions inaccessible to spray. The process overcomes the current wasteful spray techniques with almost 99 per cent coating utilization achieved. During the electrocoating operation the aluminum surface is passivated which helps adhesion and improves long term quality of the package.

Although the primary purpose in decorating the external surface of a metal container is to improve its appearance and assist its marketability, it also significantly improves the container's external corrosion resistance. Decoration of the external surface is similar in many respects to the process used to protect the internal surface, the constituents generally being dispersed in volatile solvents, applied on roller coating machines (apart from the printed image) and baked in tunnel ovens. When the container fabrication process involves severe deformation of the metal, the system used is likely to consist of a size (mainly epoxy or vinyl types), a base coat, a printed image containing several colors and a final varnish film to protect inks and base coatings from damage during fabrication and later use, and to and gloss to the print image. However, if high quality inks are used, a varnish coating may not be required.

Offset lithography has been used for over a century for decorating sheet metal. Since metal is non-absorbent, the coatings and prints cure by internal chemical reactions involving oxidation, polymerization or both. By adjusting the inks' tack properties, it is possible to print one wet ink on to the previous wet ink ("wet on wet") without the second blanket picking off the first layer. Methods used to "set" the ink in between single color presses in tandem include high-temperature air blasts, flame treatment and, more recently, the application of UV radiation to the print. The latter approach has necessitated a new approach to ink formulation. The print process is usually carried out with sheet stock prior to slitting into can body blanks or scroll shearing into end stock. With the advent of two-piece cans, less elaborate designs are used on beer and beverage cans because of the necessity to use presses that can print completely fabricated cans.

Chapter 6
The Use of Glass in Food Packaging

6.1 Composition of Glass

Glass has been defined as an amorphous inorganic product of fusion that has been cooled to a rigid condition without crystallizing.

Although glass is often regarded as a man-made material, it was formed naturally from common elements in the earth's crust long before the world was inhabited. Natural materials such as obsidian (from magma or molten igneous rock) and tektites (from meteors) have compositions and properties similar to those of synthetic glass; pumice is a naturally occurring foam glass.

Although the origin of the first synthetic glasses is lost in antiquity and legend, the first glass vessels were probably sculpted from solid blocks about 3000 B.C. In about 1000 B.C. the techniques of pouring molten glass or winding glass threads over a sand mold were developed, resulting in the formation of crude but useful glass objects. However, the real revolution in glassmaking came around 200 B.C. with the introduction of the blowing iron, a tube to which red-hot, highly malleable glass adheres. Blowing through one end of the iron causes the viscous liquid to balloon at the other end, leading to the production of hollow glass objects.

The basic ingredient of glass is silica derived from sand, flint or quartz. Silica can be melted at very high temperatures (1723°C) to form fused silica glass which has a very high melting point and is used for specialized applications including some laboratory glass.

By 200 A.D. articles of glass were in fairly common use in Roman households. During the following 1000 years, glassmaking techniques spread over Europe. However, glass remained expensive until improved techniques in the 18th and 19th centuries brought down the price of bottles and jars to a relatively affordable level.

Mechanization of glass container manufacture was introduced on a large scale in 1892 and several important developments occurred over the next few decades. These included the first fully automated machine for making bottles which was designed and built in 1903 by Michael J. Owens at the Toledo plant of Edward D. Libbey, and the development by the Hartford-Empire Company of there is (now generally taken to mean individual section, but actually named after its inventors Ingersall and Smith) blow-and-blow machine.

Further developments have occurred, resulting in the production of a wide range of glass containers for packaging. Generally, today's glass containers are lighter but stronger than their predecessors. For example, the weight of the pint (600 mL) milk bottle in England has been halved from over 567 g in the 1920s to 250 g today. Through developments such as this, the glass container continues to play a significant role in the packaging of food products.

For most glass, silica is combined with other raw materials in various proportions. Alkali fluxes (commonly sodium and potassium carbonates) lower the fusion temperature and viscosity of silica. Calcium and magnesium carbonates (limestone and dolomite) act as stabilizers, preventing the glass from dissolving in water. Other ingredients are added to give glass certain physical properties. For example, lead gives clarity and brilliance although at the expense of softness of the glass; alumina increases hardness and durability. The addition of about 6 per cent boron to form a borosilicate glass reduces the leaching of sodium (which is loosely combined with the silicon) from glass.

As a consequence of the sodium in glass being loosely combined in the silica matrix, the glass surface is subject to three forms of "corrosion": etching, leaching and weathering. Etching is characterized by alkaline attack which slowly destroys the silica network, releasing other glass components. Leaching is characterized by acid attack in which hydrogen ions exchange for alkali or other positively charged, mobile ions. The remaining glass (principally silica) usually retains its normal integrity. Weathering, although not fully understood, is not a problem in commercial glass packaging applications since it may take centuries to become apparent. However, a mild form of weathering is commonly known as surface bloom and may occur under extended storage conditions.

Glass is neither a solid nor a liquid but exists in a vitreous or glassy state in which molecular units have disordered arrangement but sufficient cohesion to produce mechanical rigidity. Although glass has many of the properties of a solid, it is really a highly viscous liquid. During cooling, glass undergoes a reversible change in viscosity, the final viscosity being so high as to make the glass rigid for all practical purposes. Although glass at ambient temperatures has the characteristics of a solid, it is a supercooled liquid and will flow even at ambient temperatures over long periods of time, albeit extremely slowly. Evidence for this can be obtained by examining very old window panes which will be slightly thicker at the bottom than at the top.

Physically, glass has a random atomic structure in that the atoms are capable of arranging themselves in different orders. The basic structural unit is the silicon-oxygen tetrahedron in which a silicon atom is tetrahedrally coordinated to four surrounding oxygen atoms. However, although the silica atoms are always surrounded by four oxygen atoms, large groupings tend to be unordered. This amorphous structure, without slip planes formed by crystal boundaries that might allow deformation, is responsible for the stiffness and brittleness of glass.

6.2 Properties of Glass

Because of its amorphous structure, glass is brittle and usually breaks because of an applied tensile strength. It is now generally accepted that fracture of glass originates at small imperfections or flaws, the large majority of which are found at the surface. A bruise or contact with any hard body will produce on the surface of glass very small cracks or checks that may be invisible to the naked eye. However, because of their extreme narrowness, they cause a concentration of stress that may be many times greater than the nominal stress at the section containing them. Because of their ductility, metals yield at such points and equalize stress before failure occurs. Since glass cannot yield, the applied stress (when it is high enough) causes these flaws to propagate. Thus it is the ultimate tensile strength of a glass surface which determines when a container will break.

In practice, a stress concentrator may be a small crack or check induced in the manufacturing process, or a scratch resulting from careless container handling. Therefore, the major step taken to make glass more break resistant involves the elimination of surface flaws (*e.g.*, microcracks) by careful handling during and after forming and annealing, since the condition of the surface has a great deal to do with its tensile properties.

The mechanical strength of a glass container is a measure of its ability to resist breaking when forces or impacts are applied. Glass deforms elastically until it breaks in direct proportion to the applied stress, the proportionality constant between the applied stress and the resulting strain being Young's modulus E. It is about 70 GPa for typical glass.

The principles of fracture analysis or diagnosis of the cause(s) of glass container breakage have been described by Moody. Four aspects are important:

Internal Pressure Resistance

This is important for bottles produced for carbonated beverages, and when the glass container is likely to be processed in boiling water or in pressurized hot water. Internal pressure produces bending stresses at various points on the outer surface of the container.

Vertical Load Strength

While glass can resist severe compression, the design of the shoulder is important in minimizing breakage during high speed filling and sealing operations.

Resistance to Impact

Two forms of impact are important: a moving container contacting a stationary object (as when a bottle is dropped), and a moving object contacting a stationary bottle (as in a filling line). In the latter situation, design features are incorporated into the sidewall to strengthen contact points. The development of surface treatments (including energy absorbing coatings) to lessen the fragility of glass when it contacts a stationary object has been very successful.

Resistance to Scratches and Abrasions

The overall strength of glass can be significantly impaired by surface damage such as scratches and abrasions. This is especially

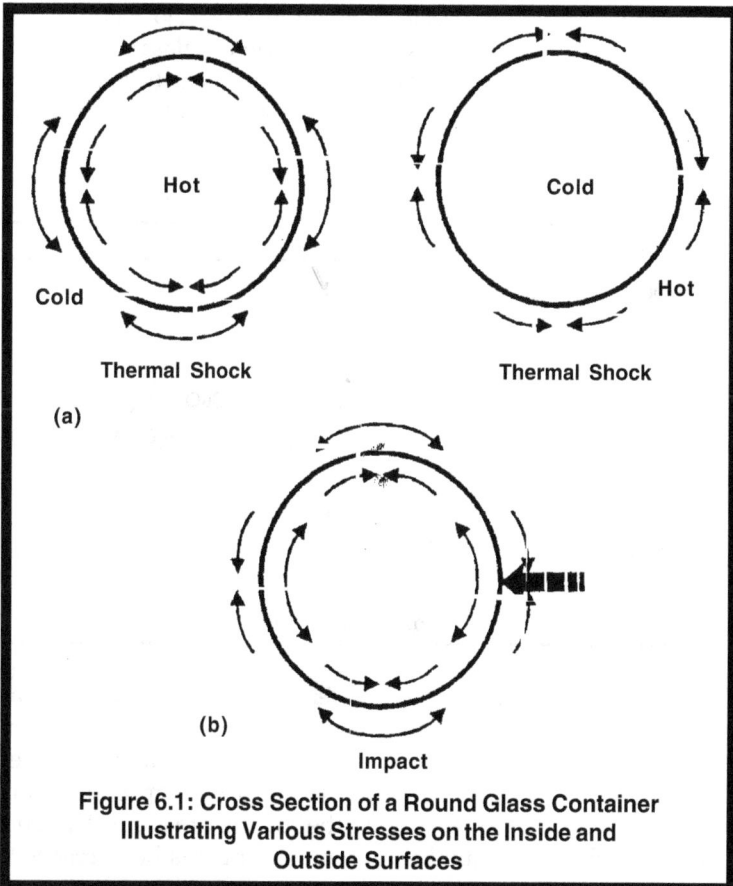

Figure 6.1: Cross Section of a Round Glass Container Illustrating Various Stresses on the Inside and Outside Surfaces

important in the case of reduced wall thickness bottles such as "one-trip" bottles. Surface treatments involving tin compounds (in conjunction with other treatments) provide scuff resistance, thereby overcoming susceptibility to early failure during bottle life.

Although the mechanical strength of a bottle or jar can increase with glass weight, this is at the expense of thermal strength which decreases with increasing glass weight. Considerable expertise is required by the glassmaker to determine the most appropriate design to satisfy the mechanical strength requirements and balance the thermal strength demands of the finished product.

The thermal strength of a bottle is a measure of the ability to withstand sudden temperature change. In the food industry, the behavior of glass with respect to temperature is of major significance, because relative to other forms of food packaging, glass has the least resistance to temperature changes. The resistance to thermal failure depends on the type of glass employed, the shape of the container, and the wall thickness.

Table 6.1: Coloring Agents Used in Glass

Effect	Oxide
Colorless, UV absorbing	CeO_2, TiO_2, Fe_2O_3
Blue	Co_3O_4, $Cu_2O + CuO$
Purple	Mn_2O_3, NiO
Green	Cr_2O_3, $Fe_2O_3 + Cr_2O_3 + CuO$, V_2O_3
Brown	MnO, $MnO + Fe_2O_3$, $TiO_2 + Fe_2O_3$, $MnO + CeO_2$
Amber	Na_2S
Yellow	CdS, $CeO_2 + TiO_2$
Orange	$CdS + Se$
Red	$CdS + Se$, Au, Cu, UO_3, Sb_2S_3
Black	Co_3O_4 (+ Mn, Ni, Fe, Cu, Cr oxides)

When a glass container is suddenly cooled (*e.g.*, on removal from a hot oven), tensile stresses are set up on the outer surfaces, and compensating compressional stresses on the inner surface. Conversely, sudden heating leads to surface compression and internal tension. In both situations, the stresses are temporary and disappear when the equilibrium temperature has been reached.

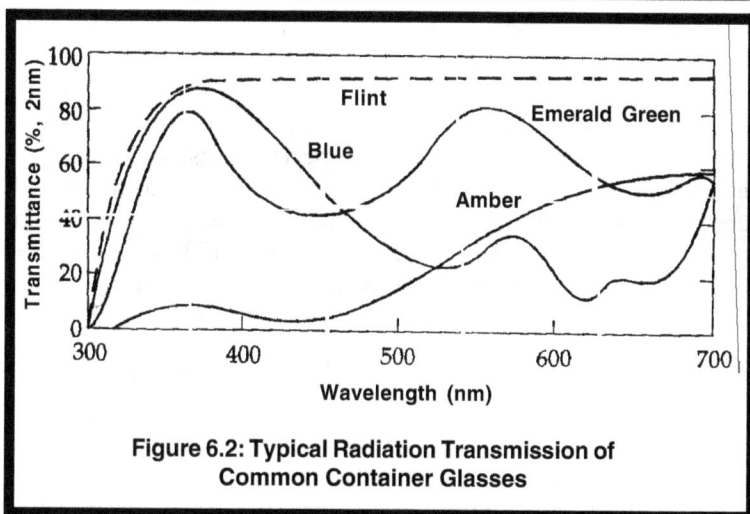

Figure 6.2: Typical Radiation Transmission of Common Container Glasses

Because glass containers fracture only in tension, the temporary stresses from sudden cooling are much more damaging than those resulting from sudden heating, since the potentially damaged outside surface is in tension. It is found in practice that the amount of tension produced in one surface of a bottle by suddenly chilling it is about twice as great as the tension produced by suddenly heating the other surface, assuming the same temperature change in both cases.

Thermal shock resistance cannot be calculated directly because the strength of glass containers is greater under momentary stress than under prolonged load. Therefore, empirical testing procedures are used. ASTM C 149 covers the determination of the relative resistance of commercial glass containers (bottles and jars) to thermal shock, and is intended to apply to all types of glass containers that are required to withstand sudden temperature changes (thermal shock) in service such as in washing, pasteurization or "hot pack" processes, or in being transferred from a warmer to a colder medium or vice versa. Resistance to breaking is determined by transferring glass containers which have been totally immersed in a hot water bath (typically at 63°C) for 5 minutes to a cold water bath (typically at 21°C) and observing the number of breakages.

Glass has no crystalline structure, so that when it is homogeneous and free from any stresses, it is optically isotropic. The optical properties of glass relate to the degree of penetration of light and the subsequent effect of that transmission, transmission being a function of wavelength. The spectral transmission of glass is determined by reflection at the glass surface and the optical absorption within the glass. In silicate glasses, transmission is limited by the absorption of silica at approximately 150 nm in the UV (ultraviolet) and at 6000 nm in the IR (infrared). Iron impurities further reduce transmission in the UV and near IR.

6.3 The Processes Used in Making Glass Containers

Bottles are normally produced by a two-step 'blow and blow' process whereby a gob of glass, accurately sheared in terms of weight and shape, is dropped into an externally aircooled cast iron cavity known as the parison or blank mold. Some of the glass flows over a plunger in the base of the mold which is used to mold the finish of the container by means of ring molds. Compressed air is applied to force the glass down onto the plunger to form the neck ring. Sometimes vacuum is applied from the bottom as an alternative or additional procedure.

When the finish molding is complete, the plunger is retracted and air blown in from the bottom, enlarging the size of the bubble until the glass is pressed out against the blank mold to form a hollow preform. This is then inverted and transferred to the blow mold where it elongates under its own weight. Air at about 200 kPa or vacuum is applied so that the glass is pressed against the metal surface of the blow mold which is air cooled to ensure rapid removal of heat. The mold is then opened and the fully blown parison removed and held over a deadplate for a brief time to allow air to flow up through the deadplate and around the container to further cool it. It is then transported to the annealing lehr.

Press and Blow (P&B)

In the case of jars, a two-step "press and blow" process is used. The body blank or "parison" is formed by pressing the gob of molten glass against the mold walls with a large plunger. When the cavity is filled, glass is then pushed down into the neck ring and the finish is formed. No baffle or counter-blow air is used in the formation of

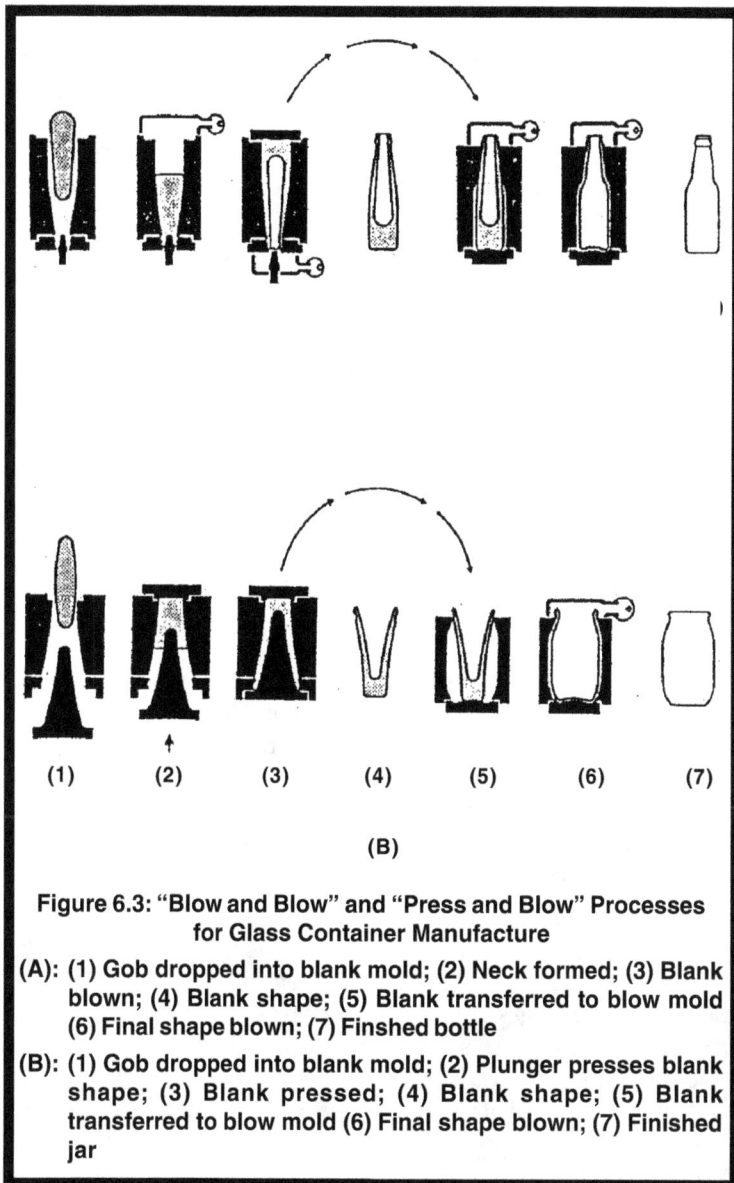

Figure 6.3: "Blow and Blow" and "Press and Blow" Processes
for Glass Container Manufacture

(A): (1) Gob dropped into blank mold; (2) Neck formed; (3) Blank blown; (4) Blank shape; (5) Blank transferred to blow mold (6) Final shape blown; (7) Finshed bottle

(B): (1) Gob dropped into blank mold; (2) Plunger presses blank shape; (3) Blank pressed; (4) Blank shape; (5) Blank transferred to blow mold (6) Final shape blown; (7) Finished jar

the parison cavity, the operation relying on the mechanical introduction of the plunger into the glass. The rest of the steps in the P&B process are identical to the steps in the B&B process.

Narrow Neck Press and Blow (NNPB)

In this relatively recent process, the gob is delivered into the blank mold and pressed by a metal plunger. The plunger and gob together have the same volume as the blank mold cavity. This enables the glass maker to decide exactly how the glass is distributed in the parison and, hence, to be able to more accurately control the uniformity of the glass distribution in the finished container. The second stage is similar to the "blow and blow" process. The parison is blown to a finished container having a more uniform wall thickness and, as a result, higher strength.

6.4 The Shape of the Glass Container

Usually the shape of the container is determined by the nature of the product, each product group having a characteristics shape. Thus, liquid products generally have small diameter finishes for easier pouring; solid products require larger finishes for filling and removing the contents. As well as filling and emptying requirements, consideration must also be given to the nature and manner of labeling the container, and its compatibility with packaging and shipping systems.

The container finish (so-called because in the early days of glass manufacturing, it was the part of the container to be fabricated last), is the part of the container that holds the cap or closure, *i.e.*, the glass surrounding the opening in the container. It must be compatible with the cap or closure and can be broadly classified by size (*i.e.*, diameter), sealing method (*e.g.*, twist cap, cork, etc.), and special features (*e.g.*snap cap, pour-out, etc.).

The finish has several specific areas including the sealing surface which may be on the top or side of the finish, or a combination of the two; the glass lug which is one of several horizontal, tapering and protruding ridges of glass around the periphery of the finish on which the closure can be secured by twisting; the continuous thread which is a spiral projecting glass ridge on the finish intended to mesh with the thread of a screw-type closure; a transfer bead which is a continuous horizontal ridge near the bottom of the finish used in transferring the container from one part of the manufacturing operation to another; a vertical neck ring seam resulting from the joining of the two parts of the neck ring, and a neck ring parting line which is a horizontal mark on the glass surface at the bottom of the

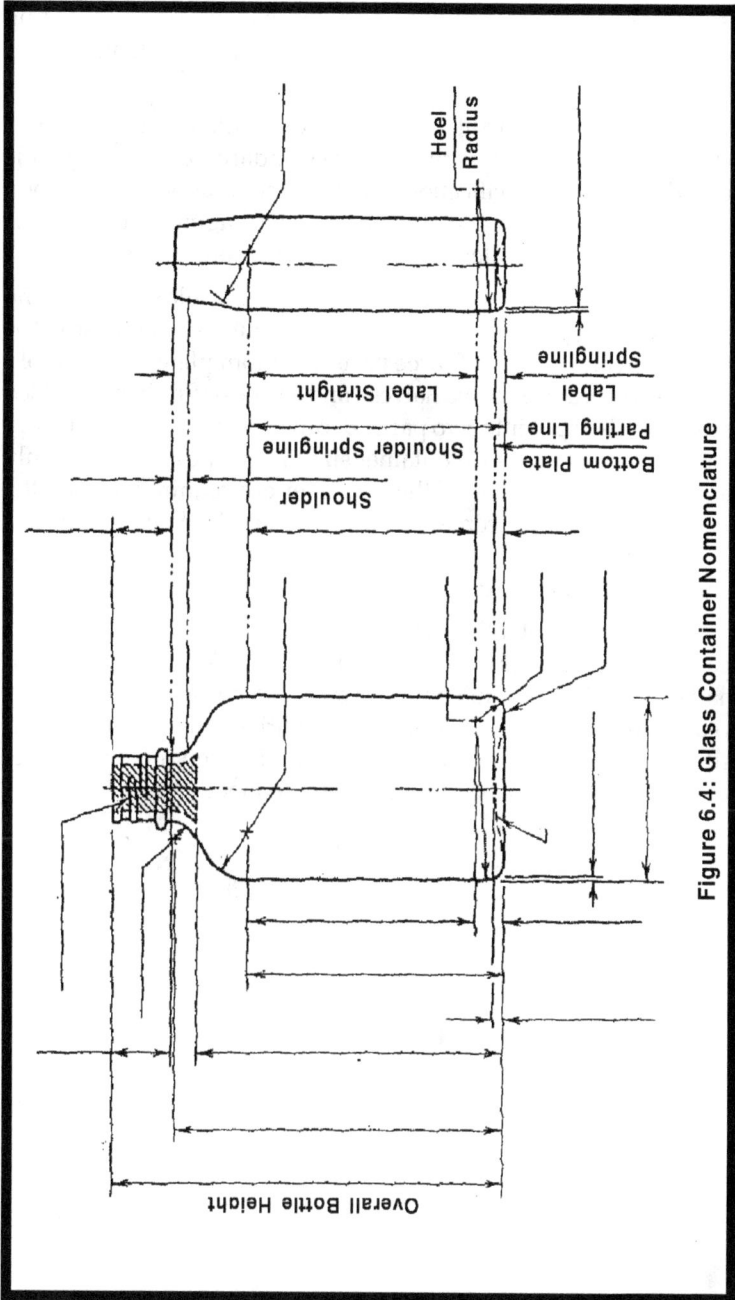

Figure 6.4: Glass Container Nomenclature

Heel Radius

Label Springline

Label Straight

Label

Parting Line

Shoulder Springline

Bottom Plate

Shoulder

Overall Bottle Height

neck ring or finish ring resulting from the matching of the neck ring parts with the body mold parts. Not all glass containers have transfer beads or vertical neck ring seams.

Although there are literally hundreds of different finishes used on glass containers, glass finishes are standardized and a specific set of dimensions, specifications and tolerances has been established for every finish designation by the Glass Packaging Institute (GPI) in the USA and equivalent bodies in other parts of the world.

Once a design has been accepted, the molds used in the manufacturing process must be made. They are usually constructed of cast iron and consist of three parts: a bottom plate, a body mold (divided vertically into two halves), and a neck or finish mold which is usually also split into two parts. Because of the high cost of mold manufacture, changes to container size and shape are usually only made if large quantities of the container are required. Generally, customers select their containers from the standard range provided by glass manufacturers unless they are extremely large users in which case the extra expense of customized designs is justified.

The GPI has established limits which are generally accepted as reasonable tolerances by most manufacturers. Allowance has been made for an increase in container size as a consequence of mold wear, as well as expected process capabilities of the manufacturer. Although closer tolerances can be met, this often incurs a higher cost since molds must be replaced more frequently.

Chapter 7

Plastic in Food Packaging: A Modern Dilemma

History

In today's world, life without plastics is incomprehensible. Every day, plastics contribute to our health and safety. The first man-made plastic was unveiled by Alexander Parkes at the 1862 Great International Exhibition in London. This material which the public dubbed Parkesine was an organic material derived from cellulose that once heated could be molded but that retained its shape when cooled. Parkes claimed that this new material could do anything rubber was capable of, but at a lower price. He had discovered something that could be transparent as well as carved into thousands of different shapes. But Parkesine soon lost its luster, when investors pulled the plug on the product due to the high cost of the raw materials needed in its production.

During the latter part of the 19th century, a rush was on to find a replacement for ivory in billiards balls. John Wesley Hyatt, an American, finally came upon the solution in 1869 with celluloid. Hyatt, upon spilling a bottle of collodion in his workshop, discovered that the material congealed into a tough, flexible film. He then produced billiard balls using collodion as a substitute for ivory. But due to its highly brittle nature, the billiard balls would shatter once they hit each other. The solution to this challenge was the addition of camphor, a derivative of the laurel tree. This addition made celluloid the first thermoplastic: a substance molded under heat and pressure into a shape it retains even after the heat and pressure have been removed. Celluloid went on to be used in the first flexible photographic film for still and motion pictures.

Cellophane was discovered by Dr. Jacques Edwin Brandenberger, a Swiss textile engineer, who came upon the idea for a clear, protective, packaging layer in 1900. Brandenberger added viscose to cloth but the end result was a brittle material that was too stiff to be of any use. Yet Brandenberger saw another potential for the viscose material. He developed a new machine that could produce viscose sheets, which he marketed as cellophane. With a few more improvements, cellophane allowed for a clear layer of packaging for any product, the first fully flexible and water-proof wrap.

The 1920s witnessed a "plastics craze", as the use of cellophane spread throughout the world. DuPont, one of the industry leaders, became a hotbed for innovation concerning plastics. Wallace Hume Carothers, a young Harvard chemist, became the head of the DuPont lab. The company was responsible for the moisture-proofing of cellophane and was well on its way to developing nylon, which at the time they named Fiber 66. During the 1940s, the world saw the use of such materials as nylon, acrylic, neoprene, SBR, polyethylene, and many more polymers take the place of natural material supplies that were becoming exhausted.

Another important plastic innovation of the time was the development of polyvinyl chloride (PVC), or vinyl. Waldo Semon, a B.F. Goodrich organic chemist, was attempting to bind rubber to metal when he stumbled across PVC. In 1933, Ralph Wiley, a Dow Chemical lab worker, accidentally discovered yet another plastic: polyvinylidene chloride (better known as SaranTM). SaranTM was first used to protect military equipment, but it was later discovered that it was great for food packaging. SaranTM would cling to almost any material - bowls, dishes, pots and even itself; thus, it became the perfect tool for maintaining the freshness of food at home.

In 1933, two organic chemists working for the Imperial Chemical Industries Research Laboratory were testing various chemicals under highly pressurized conditions. In their wildest imaginations, the two researchers E.W. Fawcett and R.O. Gibson, had no idea that the revolutionary substance they would come across - polyethylene would have an enormous impact on the world. The researchers set off a reaction between ethylene and benzaldehyde, utilizing two thousand atmospheres of internal pressure. The experiment went askew when their testing container sprang a leak and all of the pressure escaped. Upon opening the tube they were surprised to find a white, waxy substance that greatly resembled plastic. When the experiment was carefully repeated and analyzed the scientists discovered that the loss of pressure was only partly due to a leak; the greater reason was the polymerization process that had occurred leaving behind polyethylene. In 1936, Imperial Chemical Industries developed a large-volume compressor that made the production of vast quantities of polyethylene possible. This high-volume production of polyethylene actually led to some history-making events.

It was not until after the war, though, that the material became a tremendous hit with consumers and from that point on, its rise in popularity has been almost unprecedented. It became the first plastic in the United States to sell more than a billion pounds a year and it is currently the largest volume plastic in the world. Today, polyethylene is used to make such common items as soda bottles, milk jugs and grocery and dry-cleaning bags in addition to plastic food storage containers.

Plastics Packaging in Modern Life

After the Second World War, the plastics industry underwent incredibly fast development. Since the 1950s, plastics have grown into a major industry that affects all of our lives. In fact, since 1976, plastic has been the most used material in the world and was voted one of the top 100 news events of the century. There is probably no comparable sector of industry (apart perhaps from computing) which has grown so rapidly. None of the applications and innovations we take for granted would have been possible if it weren't for the early scientists who developed and refined the material. Those pioneers made it possible for us to enjoy the quality of life we do today.

The Packaging sector is the largest consumer of plastics. Around half of all goods are now packaged in plastics, and yet thanks to constant innovation and achievement of resource efficiency, these plastics account for only 20 per cent by weight of all packaging materials. The 20th Century has correctly been described as the Century of Plastics. Plastics in packaging are the perfect tool to achieve sustainable development.

Plastics Packaging Applications

Plastics, and plastics packaging, are now an essential part of our everyday life. The key to their success has been versatility. In packaging, plastics are used for many varied applications ranging from sterile storage of medical and pharmaceutical goods, to extending the shelf life of foodstuffs such as bread, meat and vegetables, and protecting sensitive technical products from damage. This means that plastics make a significant contribution to improving the quality of our life. At the same time they preserve valuable resources and help to save costs, as a result of their lower weight. Over time, plastics have become ever more sophisticated, lighter and more versatile due to innovative technologies and they have replaced traditional packaging such as glass and paper in many areas. Various applications of plastic packaging include:

1. Plastic Bags and Plastic Sacks
2. Plastic Blow Moulded Containers
3. Plastic Food Packaging
4. Plastic Packaging for Soaps
5. Plastics for Detergents and Cosmetics
6. Plastics for Pharmaceuticals
7. Plastic Films
8. Plastics for Transit Packaging

Plastics packaging industry includes food packaging, medical packaging and flexible rigid packaging. Companies in the plastics packaging industry which use their expertise to develop user-friendly new products for the market will enjoy a strategic advantage over their rivals. If they also concentrate on market niches, their long-term commercial potential will be even better.

Packages can even be produced in shapes to relate to the product and assist marketing. A current pet food industry, pack, for example, is a plastic bag for cat food, shaped like a cat. Another is a chocolate pack shaped like a rabbit. Both have dramatic point-of-sale impact and have boosted sales of the product.

There is great scope also for technical innovation. Salad goods when unpackaged have only a very short life in the supermarket. But, when they are packed in a foil bag with a protective atmosphere, their shelf life can be increased by up to ten days. A film for meat packaging, for which a new barrier layer has been developed, now allows steaks to be kept in a special controlled atmosphere for over a month.

In addition to traditional packaging, the coffee industry is now demanding packaging without an aluminium layer. Packaging manufacturers have been asked to develop a film with the same barrier properties as films that contain aluminium, which preserve the aroma of coffee for over a year. Another example is beer in flexible plastic bottles sold in football stadiums. Cold-resistant films that remain flexible and impact-resistant in a deep freeze at a temperature of minus 30°C also provide excellent opportunities. Such solutions to problems provide the basis for market-success. In development of new products, those companies offering ever-lighter and easier to recycle packaging will have a competitive edge. The reason is the growing demand from consumers and legislators for less waste, greater material savings and more environmental protection. At the same time, less material helps to preserve valuable resources and reduce transport costs. Consequently the industry is reducing packaging weight by using thinner and thinner films and thin-walled plastics containers.

There are many ways to create new applications for flexible and rigid packaging, and solve technical problems. Apart from new and modified polymers, one method is composite films, combining the benefits of individual layers to tailor-make the properties for the application. There are several processes:

Co-extrusion

Three, five or even seven layers of different polymers are combined in a molten state and extruded by the cast or the blown film process.

Adhesive Lamination

The adhesive, which may be solvent-based, water-based or solvent-free, is applied to the surface of one film.

Extrusion Coating

Molten polymer is cast-coated onto the surface of another film with a higher melting point.

Water- or Solvent-based Coating

The coating is applied to a film and the solvent or water is removed using heat.

Vacuum Coating

The latest technology is to apply a very thin layer of an inorganic material such as aluminium to a film in a vacuum chamber.

Nano-composite Technology

Nano-composite technology is also being applied to the production of films with barrier properties. By incorporation of mineral particles in nano size (less than the wavelength of light), it is possible to create a form of "labyrinth" within the structure of the film, which physically retards the passage of molecules of (for example) gas. Already, nylon films have been commercialised with improved gas barrier properties by nano-scale additives, reducing the volume of material in the package and also simplifying recycling.

Barrier properties in other package forms are achieved by co-injection and co-extrusion of blow mouldings. To improve the gas barrier of PET bottles for the vast beer packaging market, coatings of PVDC have been used, but the latest developments use plasma technology to deposit a barrier inside the bottle, integrated with the bottle production line.

Protecting the environment is also becoming increasingly established for printing the films in addition to solvent-based printing inks, water-based ones are now becoming more common.

Various Types of Packaging Plastics

Every day our lives are touched by plastic packaging products. These are some of the more common packaging products organized according to their plastic type which are as under:

PET (Polyethylene Terphthalate)

It is used in beverage containers, food containers, boil-in food pouches, processed meat packages etc. PET is more impermeable than other low-cost plastics and so is a popular material for making bottles for Coke and other "fizzy drinks," since carbonation tends to attack other plastics, and for acidic drinks such as fruit or vegetable juices. PET is also strong and abrasion resistant, and is used for making mechanical parts, food trays, and other items that have to endure abuse. PET films, trade-named "mylar," are used to make recording tape.

HDPE (High Density Polyethylene)

HDPE (High density polyethylene) is used in milk bottles, cereal box liners, detergent bottles, oil bottles, margarine tubs, toys, plastic bags etc.

PVC (Polyvinyl Chloride)

PVC (Polyvinyl chloride) is used in food wrap, vegetable oil bottles and blister packaging.

LDPE (Low Density Polyethylene)

LDPE (Low density polyethylene) used in shrink-wrap, plastic bags, garment bags, dry cleaning bags, and squeezable food bottles.

PP (Polypropylene)

PP (Polypropylene) used in margarine and yogurt containers, caps for containers, wrapping to replace cellophane, medicine bottles etc.

PS (Polystyrene)

PS (Polystyrene) used in egg cartons, fast food trays, disposable plastic silverware, cups, compact disc jackets.

To assist recycling of disposable items, the Plastic Bottle Institute of the Society of the Plastics Industry has devised the now-familiar scheme to mark plastic bottles by plastic type. A recyclable plastic container using this scheme is marked with a triangle with three "chasing arrows" inside of it, which enclose a number giving the plastic type: PETE, HDPE, PVC, LDPE, PP, PS, and OTHER.

Use of Polyolefins in Food Packaging

The safe and reliable delivery of food has emerged as one of the major industrial concerns of our time that affects both the developing and western world. Surprisingly, some 30 per cent of food losses in developing countries are the direct result of improper food packaging and failed logistics.

Compared to western methods of packaging where plastic materials are widely used and where wastes represent less than 3 per cent from farm to supermarket, these statistics are highly indicative of the strong relationship between the type of packaging material used and its ability to minimise food wastes. The global challenge of food delivery has thus forced recognition of the importance of materials used for food packaging. Indeed, packaging materials must live up to tough requirements, safely preserving and protecting food contents without compromising the health of consumers. The materials must equally be viable commercial solutions, providing a unit cost reduction and delivering shelf appeal that is attractive to the end user.

Although the packaging industry can choose from an array of materials for food packaging, polyolefins (polyethylene and polypropylene) present a flexible and robust solution for food packaging, offering a balance between performance, processability and cost to make them the industry's material of choice. Polyolefins is the first choice of material for advanced rigid packaging.

One of the most visible parts of the plastics invasion was Earl Tupper's "tupperware," a complete line of sealable polyethylene food containers that Tupper cleverly promoted through a network of housewives who sold Tupperware as a means of bringing in some money. The Tupperware line of products was well thought out and highly effective, greatly reducing spoilage of foods in storage. Thin-film "plastic wrap" that could be purchased in rolls also helped keep food fresh.

Polyethylene and polypropylene often offer the lowest unit cost of any packaging system when combined with an optimised pack design. Their low densities and high performance levels, made possible with state-of-the-art high performance grades, offer extreme light weighting.

Whether considering a full life cycle analysis or the packaging conversion process alone, polyethylene and polypropylene are materials that favourably address environmental concerns such as energy consumption, climate protection, and water conservation, compared to competing and traditional packaging materials.

Polyethylene and polypropylene intrinsically offer low densities which mean low packaging unit weight. This results in an important reduction of material used and minimises waste at end of life. The materials also significantly lower fuel consumption for transport and contributes to lowering our CO_2 emissions. In addition, advanced polyolefins are recyclable or can be turned back into energy to substitute fossil fuels in clean energy recovery processes.

Plastic Film

Plastic is any material made of polymeric organic compounds and additives that can be shaped by flow. This is the most common material used to make high quality and highly significant films.

In 1868 John Hyatt mixed celluloid with camphor and alcohol to find a substitute of billiard balls. Then came celluloid collars, cuffs, etc. After the turn of the century, cellophane was invented. The major growth for cellophane and all plastic films started during the late 1930s and after World War II when self-service shopping came into vogue. Many packages required transparency because people wanted to see what they were buying. Since then, many plastic films were added to the list of packaging films.

Plastic films are high performance materials, which play an essential part in modern life. They are mostly used in packaging applications along with some applications in agricultural, medical and engineering fields. Plastic films are perfectly stable, easy to work with and can be lighter than tissue.

There are many material types used in plastic films ranging from single layer polymers to multilayer polymers with tie layers and copolymers. A number of plastics are used in applications such as nylon, polypropylene, cellophane, etc.

Polypropylene Films

These are heat sealable films for flexible packing and have high tensile strength. The film can contain colorants, stabilizers, or other additives, and can be coated for the improvement of performance properties.

Polyester Films

Polyester film features superior performance in printing, metalizing and cold seal applications. The film shows excellent scuff resistance, machinability and temperature characteristics. The adhesive can be clear or pigmented for ease of identification.

Cellophane Films

Cellophane film is made by the viscose process, a physical process for making regenerated rayon by treating cellulose with caustic soda, and with carbon disulfide to form cellulose xanthate, which can then be spun into fibers and reconverted to cellulose by an acid treatment.

Nylon Films

Nylon film is a transparent, non-heat stabilized plastic film. They absorb water. The higher the moisture content the more flexible they become. They offer excellent thermal and chemical resistance, with high tensile strength, and tear resistance.

Plastic films do not have voids in the surface that allow an ink or coating to penetrate into it. Certain plastics can be attacked or swelled by specific UV raw materials. These can be matched up with the type of substrate and allow the ink to penetrate the film.

Applications of Plastic Films

Plastics are considered to be one of the most valuable and versatile family of materials ever developed. New uses are developed for plastic films almost everyday. Plastic film offers a number of significant benefits:

1. Plastic food wrappings prevent spoilage.
2. Plastic wrappings reduce food waste.
3. Plastic vapour and air barriers prevent moisture damage which in turn helps in supporting energy-conservation efforts in home construction.

4. Plastic stretch wrap and shrink wrap reduce packaging weight and bulk. This helps in reducing transportation and storage costs.

5. Agricultural plastic film helps in reducing weeds, keeping seedlings moist, wrap silage and increase crop yields. From the time these films are produced, to their use and reuse, and to their final recycling, or disposal, plastic films provide environmental benefits.

6. Plastic film can be coloured or clear, printed or plain, or laminated to aluminum, paper and other materials.

Market applications for plastic film can be divided into the following categories: Plastic film offers several packaging options. The films can be used to produce:

Food Packaging

Plastic films for packaging are available in the the form of packaging pouch, packaging bags, packaging rolls, sheets, foils etc. Food packaging includes bags for bread and rolls, in-store bags for produce and bulk foods, candy wrap and bags, bag-in-a-box, carton liners for cereal and cake mixes, wrappers for fresh food, and wrappers for prepared red meat, poultry and fish, milk bags, grocery bags.

Non-food Packaging

This include industrial liners which can be used for everything from tote boxes to large drums to bubble packaging, shipping sacs, mailing envelopes, dry cleaning bags, diaper overwrap, stretch wrap and more.

Examples of Innovations for Advanced Packaging

Source Reduction/Light Weighting in Thin Wall Consumer Packaging for Yellow-fats, Dairy and Similar Markets

It is a new high stiffness packaging with high impact polypropylene block copolymer. The exceptional flow and strength allows processors to use injection moulding to achieve packaging with walls as thin as 0.3 mm without compromising on impact resistance.

Aesthetics/Appearance Improvements in Ice Cream Packaging

It is a transparent polypropylene for proven consumer appeal with impact properties designed for deep freeze storage and transportation at -20°C.

Processing Efficiency with Cycle Time Reduction in Pail Production

The high flow polypropylene block copolymer with high impact allows fast injection cycle time in multiple cavities and in more demanding shapes *e.g.* rectangular paint pails. This allows unit cost reductions in both machine time and energy utilisation.

Convenience *e.g.* Easy to Open in Sports-lock Bottle Closures

The design freedom of a random copolymer polypropylene allowing convenience features in Sports-lock' type bottle closures but with the outstanding taste and odour properties previously only associated with the mould design limited 'Organoleptic HDPE' grades.

Hot Fill Capability for Juice, Dairy, Sports-drink and Other Liquid Packaging

The market leading injection stretch blow moulding polypropylene (ISBM PP) grade that offers outstanding transparency, high productivity and light weighting capability, while also allowing improved hot fill capability versus competing HDPE and PET bottle solutions.

Plastic Packaging: A 'Service' Industry

If a packaging manufacturer offers complete solutions, the customer can transfer entire responsibility for the production process to the supplier. This outsourcing opportunity is increasingly used by customers.

It also includes innovative advice on packaging problems and reliable, fast deliveries, up to 'Just-In-Time' supply. The packaging industry is looking to cooperate more closely with suppliers of raw materials and machinery.

The packaging industry, raw material suppliers and machinery manufacturers are already cooperating globally in Europe, USA and in Japan. Fast methods of communication by e-mail and computerised design ensure that this cooperation is even more effective.

For example, the food industry needs re-sealable peelable packaging allowing products such as vegetables and meat to be removed, resealed and removed again later.

Laminates are another example of inter-sector cooperation, and here the packaging industry is working with manufacturers of adhesives and machinery to develop eco-friendly solvent-free laminates with a shorter curing time.

In other sectors, the potential of PET for high-quality clear rigid containers is still only at its beginnings: wide-mouth, hot-fill, high barrier solutions are moving from laboratories to production lines.

Advantages of Plastic Usage in Food Packaging

When the packaging questions are tough, plastics are often the answer. Sometimes they are the only answer, performing tasks no other materials can perform and providing consumers with products and services no other materials can provide.

Plastic packaging continues to have the wrap on consumer preference. Freshness, storage stability and ease of preparation are among the consumer goals driving the popularity of plastic food packaging. Offering safety, quality, convenience and savings, plastic packaging meets the needs of consumers. Different plastics offer different qualities, giving manufacturers and consumers the freedom to choose the type of plastic that best suits the application. Plastics can be rigid when protection is needed, or flexible for convenience's sake. They can be clear or opaque. And they can be molded into a wide variety of shapes and sizes. New ideas include plastic containers for cereal, coffee, spices and baby food, as well as squeeze bottles that allow portion control of juice concentrates and keep contents fresh in the refrigerator for up to five weeks. Freezer-to-oven-to-table plastic food packaging is now available for both microwave and conventional oven use. And plastic container design itself is participating in the cooking, with innovations such as tapered popcorn boxes that keep the kernels in the hot oil and microwavable cake mixes with reusable trays.

In other types of packaging as well, consumers and the hospitals, schools and other institutions that serve them increasingly are turning to plastics. Safe, sanitary, easy to use and economical, plastic packaging is the shape of the future. Various advantages of plastic packaging are:

Safety by Design

In the home, break-resistant, shatterproof and no-spill plastic bottles cut down on injuries and clean-ups anywhere the floor is hard and hands may be slippery. Food-service outlets and their customers rely on plastic packaging to protect food products against contamination and retain desired temperatures longer. And single-serve plastic packaging for condiments not only preserves freshness and flavor, it also ensures the consumer a sanitary portion while cutting down on food waste.

Plastic packaging molds itself to modern lifestyles. Today's working parents and busy homes rely on its convenience and the services it provides. Microwave ovens have become a near necessity in homes, and plastic trays are the package of choice for consumers. Microwave cooking enables active people to eat well without spending their limited leisure time on food preparation. The elderly, too, benefit from the ease of microwavable food packaging.

Plastic packaging also preserves flavor and saves time in conventional cooking and storage. Squeeze bottles for condiments, boil-in-bag dishes, resealable bags for everything from shredded cheese to cereal, freezer bags that protect food against ice crystals, precooked foods that are microwavable in the package all contribute to quality meals in the home.

Extremely lightweight and molded to promote easy handling, plastic containers allow consumers to enjoy the savings of beverages, detergents and other products in the "large economy size." And plastic packaging, which can be transparent without being fragile, enables consumers to see what they're getting and to serve themselves.

In addition to saving space in today's smaller living quarters, plastic packaging can be as decorative as it is serviceable. Further, it won't leave rust rings on counters and fixtures. For a host of personal and home products, plastic packaging works well and looks good, too.

Differentiation through Design

Plastic packaging is unique in that it allows for a variety of design capabilities. Design innovation resulting in both aesthetic and functional value helps to differentiate end-market products on the shelf. Due to the large number of end-market products on the shelves today, design elements have become an increasingly important distinguishing characteristic in the eyes of the consumer.

A major trend in packaging design is the use of shrink-sleeve labeling. Shrink-sleeves are most often produced with PVC, but can also be produced from PETG (Polyethylene Terephthalate Glycol Co-monomer) and OPS (Oriented Polystyrene), which have different shrink characteristics and may be more appropriate for certain applications. Using this type of labeling provides 360-degree design capabilities that allow for maximum utilization of the container surface for optimal graphic visibility. In markets where product differentiation is limited, shrink-sleeves can offer a highend, attention-grabbing design that distinguishes the product from others on the shelf.

Packaging that enhances functionality of a product can also have an impressionable effect on the consumer, and plastic holds an advantage over some substitute materials, with its ability to be formed into any number of shapes that compliment or improve product usage. The production of dual-chamber containers, where two products mix as they are dispensed, is one example of a product design gaining popularity. In addition, packaging attributes that address the needs of on-the-go lifestyles such as single-use pouches, handles for easy carrying, and re-closable openings will continue to grow.

Use for Shipping, Storing, Savings

The use of plastics for shipping and storage will continue to grow. Strong, durable and tear resistant, plastic packaging saves energy, space and money. Plastic containers, which generally require less energy to manufacture than other packaging, also require less fuel to transport than heavier materials. Additional savings come from reductions in shipping damage and elimination of the need for additional packing materials, such as partitions between individual products. Strong enough for stacking and moldable into space-saving shapes, plastic containers can maximize warehousing room and lower storage costs.

Meeting unique packaging needs from anti-static protective packaging for electronic components to shelf-stable containers for products that once required costly cold storage are a specialty of plastics. Because they can be molded to fit contours, plastics provide the ultimate protection in packaging office machines, entertainment units, food products and other delicate products. Tough enough to withstand the stresses of transportation yet capable of screening out even the smallest particle of dust, plastic packaging delivers. These factors all add up to savings for producers and merchants and can result in lower prices for consumers.

Convenience

Consumer lifestyle changes have recently resulted in an increased emphasis on convenient, on-the-go packaging options, especially in the food and beverage industries. Characteristics which cater to these preferences include light weight, portability, and durability. Now that plastic has achieved a number of protective capabilities on par with glass and metal substitutes, it has begun to differentiate itself as the best option for new product innovation due to its wide range of applications.

Flexible plastic packages, such as stand-up pouches, have recently been gaining market share at the expense of some rigid packaging alternatives. Similar to films, less resin is used in manufacturing, which lowers production costs. In addition, flexible packaging requires significantly lower transportation costs and less customer shelf space than rigid packaging. Pouches also offer lightweight portability, effective single-serving options, and extensive marketing capabilities. Pouch packaging is not a new technology; in fact, it has been used since the 1960's by the U.S. Army. However, improved barrier properties, combined with current consumer life-styles and a growing demand for ready-to-eat (RTE) products have resulted in a surge of demand. Retort pouches are popular for these kinds of applications. In the retort process, the pouch is filled, sealed, and then retorted through a thermal process that cooks or sterilizes the product. Plastic pouches now have shelf-life comparable to that of metal or glass containers, and also have the advantage of being suitable for microwave preparation, which caters to a convenience driven consumer base. Companies who can effectively capitalize on the advantages of pouch packaging in the near future are expected to capture market share in this highgrowth space.

Barrier Protection

Recent trends in the plastic packaging industry have led flexible packaging producers towards the manufacture of thinner gauge films due to the material reduction advantages. Product pricing is an important competitive factor because most of these products are high volume, low margin items, which has led to commoditization and a lack of differentiation in most film, sheet, and bag. With resin costs accounting for approximately two-thirds of the cost of goods sold, keeping raw material costs to a minimum has become a priority.

Coinciding with the push towards thinner gauge plastics is increasing customer demand for improved performance and barrier protection. In addition, products used to package foods, beverages, and pharmaceuticals need to comply with legislative requirements of state and federal health and safety authorities. The food and beverage industry is focused on manufacturing thinner gauge film with high barrier protection, and the pharmaceutical industry is determined to realize continued improvement of child-resistant and tamper evident closures. Producing packaging that provides for better performance and protection is a major objective driving innovation in the space.

Flexible films can be produced with very little resin, which reduces production costs when compared to other packaging materials. Co-extrusion technologies have enabled manufacturers to combine various resins in order to achieve a highly effective barrier, while at the same time producing thinner gauge films that cost less to the manufacturer. However, combining two or more materials in the package has created recycling problems for many co-extruded films.

Two specific packaging technologies that have been recently gaining momentum are modified atmosphere packaging (MAP) and nanotechnology. The MAP technique involves the practice of altering the composition of the internal atmosphere of a package in order to improve a product's shelf life. This trend is most common in foods, but is also used in pharmaceutical packaging. Nano-composites are plastics that have fillers dispersed throughout the resin which reduces permeability in the resulting film. Nanotechnology offers enhanced barrier protection, increased shelf life, and light-weight design. These developments have significant potential in improving the barrier performance of plastic packaging films and creating new markets for plastic packaging.

Globalization of Plastics Packaging Industry-Economic Factor

Plastics packaging came into widespread use with the introduction of polyethylene in the fifties. Before this, in addition to classical packaging materials such as paper, glass and wood, we used films of converted natural materials such as cellulose acetate and cellophane transparent cellulose film. The development of polystyrene, polypropylene, PVC, polyesters and polyethylene copolymers saw the start of the rapid increase in the use of plastics.

Plastics packaging is everywhere today. In spite of the size and economic importance of the industry, it is a fact that manufacturers of plastics packaging (predominantly medium-sized and small companies) are now open to a double dependency: on the one side the raw material suppliers dictate the prices of plastics, and on the other side there is massive downward pressure on prices by customers - particularly in the food industry.

In addition there is increasing competition, especially from Eastern Europe, where manufacturers have capacities for high quality extrusion and printing at a lower cost. One can also expect increasing competition from the Far East, although manufacturers there are yet not as advanced as in Europe, particularly in the fields of barrier materials and printing technology. An additional pressure is that the packaging market is characterised by growing overcapacity, which is placing even more pressure on prices. Nevertheless demand for packaging will continue to increase.

It is anticipated that growth rates over the last few years averaging 4-5 per cent per annum will remain at these levels, or increase. There also appears to be a slight trend in Europe towards flexible packaging systems.

As far as the individual plastic materials are concerned, very high growth rates are expected for PET, in the field of rigid packaging. The main application for PET is bottles for carbonated drinks and mineral water. In Europe and the USA, PET bottles are now also being tested for packaging beer.

Another application for PET which will gain in significance is food (such as preserves) that is hot-filled. Growth is also expected in the use of injection moulded polypropylene for large tubs and buckets that will gradually replace metal containers. PVC, which is under pressure from other materials in the food sectors, holds a strong position in 'bubble' packaging of pharmaceutical tablets and in 'display' packaging of products such as tools, ironmongery and hardware.

In terms of product areas, food packaging which is the largest single product area in the whole packaging industry will be the major growth market for plastics packaging. The growth of the market is assisted by demographic developments in the modern world, such as the steady increase in single and two-person households and the growing number of elderly people, which is fundamentally influencing consumer purchasing habits.

The market demand today is for practical, time-saving ready or deep-frozen meals in small microwaveable easy-to-open packaging or packaging that can be resealed. The trend towards these convenience products is enhanced by the fact that more and more people can now afford them. The flexibility of plastics, in protective properties and processability, has given them an excellent position for fulfilling the specific packaging requirements of this market.

Another new and growing special market is the packaging of pharmaceutical and medical products. Again, there is a strong and growing demand for these products in industrialised countries, where considerable consumer expenditure is available for healthcare. These products, such as medicines, prostheses and hygienic products place particularly high demands on packaging, for sterility, protection, appearance and security, which can well be met by plastics.

The increasing level of the globalisation of business processes has led to structural changes for raw material suppliers and customers of the packaging, which have far-reaching effects on plastics packaging companies. As for all large companies, it is essential for large purchasers of packaging today to have a global presence. They expect that packaging manufacturers will be capable of providing local deliveries to them, wherever they may be located. Those who manufacture products in China assume that their packaging supplier will also build a factory there.

Increasingly raw materials suppliers, such as BASF and Shell, are merging in full or part to remain competitive in the global market. The trend is moving towards raw materials supplies being restricted to pure plastics without any additives, in order to streamline production. Plastics processors will need to purchase additives such as stabilisers, pigments and lubricants separately and then mix them themselves to obtain the correct blend.

This means that the raw material suppliers will shift responsibility for raw materials to the plastics processing industry. If this trend becomes established it will be essential that the plastics processors acquire the necessary expertise and invest in new specialised machinery to meet this challenge. It is also unavoidable that small and medium-sized companies will have to join forces to form larger groups and operate on a global level.

This means that, on the one hand, they will form a counterbalance to the raw material suppliers, while on the other they will only be able to offer their customers a comprehensive range of packaging through consolidated purchasing with other plastics processors.

At the same time, wide-ranging joint ventures between companies will allow small and medium-sized companies to gain better financial control. On a political level, too, co-operation is urgently required. There is a whole series of very professional national and sector associations, such as the Plastic Packaging Industry Association in Germany, the British Plastics Federation and La Federation de la Plasturgie in France, to name but a few.

It is important that the industry speaks with one voice to ensure that it does not splinter. Already there is fruitful cooperation at European level with the European Plastics Converters Federation. This activity needs to be intensified and is particularly important in the fields of environmental protection and recycling. The European Commission together with the Packaging and Packaging Waste Directive of December 1994 imposed far-reaching regulations, including those for the reduction of packaging material and a high level of recycling. In Germany the packaging industry has suffered a loss of competitiveness on a European level as a result of the expensive waste separation system of the Dual System. Regulations need to be standardised in all countries to prevent this imbalance.

Plastics in the Environment

There is a growing awareness of the health and environmental consequences of food packaging, especially plastic packaging made from petroleum products. There are multiple costs to our increasing use of packaging for food and a conversation about this use is probably in order.

Reduction of the use of packaged and throw-away items seems the right course. Reitman-White and Doppelt (2006) are correct when they say that retailers like PCC and The Food Co-op can't change the market by themselves. Consumers also need to take responsibility for their own personal choices, by choosing more sustainable products, and by simply choosing to use fewer packaged products.

A Throw-away Society

There's no dispute that the Western World has a throw-away problem. According to the Environmental Protection Agency, municipal waste has increased more than 50 per cent since 1980 to its 2003 level of 236 million tons per year in USA. Of that total, 131 million tons of garbage go straight to the landfill, 33 million tons are incinerated producing air pollution, and a scant 72 million tons are diverted into recycling or composting (including yard waste). The average American throws away 4.5 pounds of garbage every day, of which only 1.4 pounds is recycled or composted.

Throw-away Facts

1. Consumers and industry throw away enough aluminum to rebuild our entire commercial air fleet every three months.

2. Each recycled aluminum can saves the energy equivalent of half a gallon of gas.

3. Americans go through 2.5 million plastic bottles every hour.

(Source: *The Derrick News-Herald*, October 30, 1998)

All this packaging contributes not only to the landfill, but also consumes a huge amount of resources including energy to produce. Despite a growing amount of postconsumer content, most paper products are made from trees cut from our dwindling forests. Plastics take up 6 to 8 per cent of the total oil and gas we use, a portion that will look more and more significant as we exhaust the world's oil and gas supplies. In addition to environmental concerns, food packaging raises health concerns. Many studies warn that potentially carcinogenic and hormone-disrupting chemicals in plastic food packaging can leach into food. Leaching increases when plastic comes in contact with oily or fatty foods, especially when heated, and from old or scratched plastic. Types of plastics shown to leach toxic chemicals are plycarbonate, PVC and styrene. A growing number of environmentally concerned consumers are creating a hot market for new, biobased products. Could these bio-plastics replace oil-based plastics?

PVC: The Toxic Plastic

Polyvinyl chloride, also known as vinyl or PVC, poses risks to both the environment and human health. PVC is also the least recyclable plastic.

1. Vinyl chloride workers have elevated risk of liver cancer.

2. Vinyl chloride manufacturing creates air and water pollution near the factories, often located in low income neighborhoods.

3. PVC needs additives and stabilizers to make it unsealable. For example, lead is often added for strength, while plasticizers are added for flexibility. These toxic additives contribute to further pollution and human exposure.

4. Dioxin in air emission from PVC manufacturing and disposal or from incineration of PVC products settles on grasslands and accumulates in meat and dairy products and ultimately in human tissue. Dioxin is a known carcinogen. Low level exposures are associated with decreased birth weight, learning and behavioural problems in children, suppressed immune function and disruption of hormones in the body.

Plastic as Green Packaging

Environmental concerns influence the packaging industry, as indicated by increasing pressure on regulatory agencies from lobbyist groups. Standards for recycling are becoming more stringent, and pressures to reduce the amount of material in packaging are mounting.

This pressure is an additional factor that is causing a shift away from rigid packaging in favor of flexible options. Flexible packaging uses less material for the same application, and frequently the resins used in flexible packaging are more easily recycled. A plastic pouch will typically use 50 per cent less material than a rigid plastic container or bottle for the same application, and up to 75 per cent less material (by volume) than a glass container or bottle.

Bioplastics are also beginning to make headway in the plastics packaging space as biopolymers have recently begun to gain significant attention industry-wide after years of undelivered promises surrounding biodegradable and eco-friendly materials. One of the most influential biopolymers currently marketed is NatureWorks' corn-based polylactic acid (PLA.). Major corporations and venture capitalists are banking on it. NatureWorks LLC, a subsidiary of Cargill Dow, opened a new factory in Nebraska in 2001 and is generating more than 300-million pounds of corn-based plastics per year using 40,000 bushels of Cargill corn each day in the process. This corn-derived polymer is used for soft-drink cups, salad and fruit containers, as fill in pillows and comforters, and even in donut boxes and gift card wraps. But is it really an environmentally friendly alternative?

PLA offers packaging manufacturers a cost-competitive, renewable option to replace traditional petroleum-based plastic materials. PLA uses 68 per cent fewer fossil fuel resources than traditional plastics in its manufacturing and is the world's first greenhouse-gas-neutral polymer. In addition, current research shows NatureWorks PLA can exist in the present North American infrastructure with the existing commercial systems for recycling PET and HDPE.

Spartech Corp. has recently announced the production of Rejeven8, made from 95 per cent NatureWorks PLA and is said to better match properties of traditional PET. Some manufacturers are already beginning to use PLA in thermoformed and shrink wrap packaging applications. By doing so, retailers are able to differentiate themselves on the shelves, by catering to an increasingly eco-friendly consumer base.

Packaging and Plastics: What's a Consumer to Do?

One of the most common and persistent concerns raised by PCC members is the need to reduce packaging, especially plastic packaging. Sustainable solutions seem elusive, though new possibilities are emerging. How can conscientious consumers support a market for more sustainable products? How do we distinguish between real, long-term solutions and short-term fixes that appear attractive but fail in the long run to protect our environment and health? One such conundrum involves the new biodegradable, plant-based plastics. Will these be the wave of the future? Informed consumers increasingly are concerned about the environmental and health risks of packaging particularly the packaging of our food.

Compostability and Recycling Issues

Advocates highlight that corn-based plastics are 100 percent biodegradable as long as they're in a controlled environment with at least 140 degree temperatures and 90 per cent humidity for 40 to 60 days. Given these conditions, they degrade into carbon, hydrogen and oxygen.

Unfortunately, most consumers don't have access to the kind of industrial composting facilities needed to create these conditions. One composting specialist says few people know that corn-based plastic requires very specific conditions to break down; it will not degrade in backyard compost, nor will it degrade more quickly than other plastics in a landfill. Another common misperception is that corn-based plastic can go into residential recycling bins with other plastics. This is not true. There currently is not a large enough market to recycle this type of plastic.

Some professional recyclers express concern that PLA can contaminate the #1 plastic recycling stream because it's indistinguishable from other clear plastics and may be mixed in mistakenly. Others think recycling is unlikely to be affected adversely until PLA makes up a far greater percentage of the plastic stream.

GMO Issues

Even if consumers can get their corn-based plastic to a composting facility, many eco and health-conscious consumers still are critical of the raw material used to create the product. Corn-based plastics are made from non-organic corn, which already is at least 46 per cent of the U.S. corn planted. GE corn is so widespread that many companies, such as Patagonia and PCC, have chosen not to use corn-based plastics until a non-GE source of raw material can be guaranteed. Others have opted for the Source-Offset program offered by NatureWorks, whereby, for every pound of NatureWorks PLA delivered, a corresponding amount of non-GE field corn is purchased. Theoretically, NatureWorks PLA could be made from other high-sugar crops such as sugar beets (also genetically engineered). But the only economically viable raw material today is corn.

GE-corn Issues

GE corn is the most petroleum intensive crop except for cotton. Petroleum is used to power machines in cultivation and harvest, to make synthetic fertilizers and pesticides, to transport corn for manufacture into bioplastic, and to shred and heat the bio-plastic for composting. GE corn threatens organic farmers and a sustainable environment. Pollen drift and genetic contamination reduce biodiversity. Insects develop resistance to the engineered Bt toxin over time, threatening to render natural Bt sprays useless for organic farmers, a tool allowed as a last resort. GE corn kills the larvae of beneficial pollinators, including the Monarch butterfly, and other beneficial insects. It fails, for example, to kill the African Cotton Worm, but kills the beneficial Lacewing that eats Cotton Worms.

Why not GE Corn Plastic?

A society wanting sustainable solutions needs to avoid quick fixes that aren't sustainable and ultimately make things worse. From a systems perspective, GE corn plastic is a classic case of a "fix that fails" and "shifting the burden" dynamics.

GE corn plastic builds a market and provides justification to expand GE acreage. Expansion is certain to have negative economic, social and environmental consequences for PCC and many others. Crop diversity slowly but surely will be reduced, putting the entire food system at risk. Large industrial agricultural corporations will exert even greater control over production and distribution, which increasingly will threaten small farmers and organic food production. This will lead to increased prices because there'll be fewer suppliers controlling the market. In short, use of the product now may help in the short run, but harm PCC's business in the future.

While GE corn plastic might seem to make sense in the short term, its many negative consequences could make it harder or even impossible ever to return to sustainable solutions. Bottom line: it's no solution to replace one problem with another. Better to avoid GE bioplastics now and encourage better choices that won't harm people, the economy and the environment for a long time into the future. With this type of issue, "slower is faster and faster is slower."

Resource Conservation

Conserving resources means using less raw materials and energy throughout a product's life cycle from development through disposal or recycling. Plastics are derived from natural resources typically oil and natural gas. And yet, in part because of plastics' unique characteristics *i.e.* lightweight, durability, formability etc they can conserve more resources during a product's life when compared to some other materials.

After their intended use, plastic containers often can be used again for the same or a different purpose. Plastic grocery sacks can tote wet swimsuits home from the beach or garbage to the bin. In the hands of ingenious consumers, plastic milk jugs become planters and plastic soda bottles are converted to bird feeders.

In 1994, over 1 billion pounds of plastics were recycled. That figure has grown dramatically as new technologies, markets and collection systems are developed. Some 15,000 communities have more than tripled their collection of plastics for recycling in the last four years. Recycled plastic soft-drink bottles are being used to create new products, such as new bottles, fiberfill for winter clothing, carpeting and building materials. Recycled plastic milk jugs and soft-drink-bottle base cups are being used to create drainage pipes, buckets and plastic "lumber" for boat docks. And mixed plastics are being recycled into landscaping "timber" and outdoor benches.

Plastic packaging, which constitutes less than 4 per cent of all municipal solid waste by weight, also can be disposed of safely in landfills. And when incinerated, plastics - with their high energy content help the waste mix burn more efficiently, enhancing waste-to-energy conversion and leaving less ash for disposal.

Plastic Recycling and Beyond

The plastics industry supports recycling that is sustainable, economical and environmentally responsible. Since 1990, the plastics industry, as individual companies and through organizations such as The American Chemistry Council's (ACC) Plastics Division, has invested more than $1 billion to support increased recycling and educate communities in the United States.

Plastic Bottles

When it comes to resource conservation, recycling is one area where consumers can make a big difference. Lightweight, shatterproof plastic beverage bottles are great to take to work, on errands or just about anywhere. They are also among the most readily recycled plastics. In 2005, plastic bottle recycling reached a record high of more than 2.1 billion pounds in the United States, and the overall plastic bottle recycling rate climbed to 24.3 per cent. While this is all good news, there is a tremendous opportunity to recycle even more namely, the remaining 75.7 per cent of plastic bottles used in this country annually.

Before one pitches that empty water bottle into the trash, consider this: Every ton of plastic bottles recycled saves about 3.8 barrels of oil. Recycled plastic bottles are used to make hundreds of everyday products, ranging from fleece jackets and carpeting to detergent bottles and lumber for outdoor decking. Over 80 per cent of U.S. households have access to a plastics recycling program, be it curb side collection or community drop-off centers, yet, in most areas, the demand for recycled plastics exceeds the available supply. In recent years, the number of U.S. plastics recycling businesses has nearly tripled. Today, more than 1,600 businesses are involved in recycling post-consumer plastics.

Plastic Bags

From carrying groceries to taking out the trash, plastic bags are used and reused for dozens of tasks every day. In addition to being convenient and recyclable, plastic bags have many positive attributes from an environmental perspective:

Energy Efficient

Plastic grocery bags require 20-40 per cent less energy to manufacture than paper ones.

Lightweight and Compact

Lightweight plastic bags help save space and fuel in transport. For every seven trucks needed to deliver paper bags, only one truck is needed for the same number of plastic bags, helping to save energy and reduce emissions.

Reusable

More than 80 per cent of today's consumers reuse plastic bags as liners for household wastebaskets, shoe totes and laundry or garment bags.

Recyclable

Recycled plastic bags can be made into a wide range of second generation products, including new bags, exterior house siding, and plastic lumber for decking, park benches and picnic tables.

High Fuel Value

Plastics can help trash burn more efficiently in energy-recovery facilities, creating energy that can be used to make electricity in some communities.

Plastics Disposal

Despite the continuing growth of recycling, improvements in source reduction and increases in energy recovery, some waste will always require disposal. One should learn about plastics in landfills and their important role in protecting our environment and our groundwater from potential hazards posed by landfills.

Chapter 8

Thermoplastic Polymers

8.1 Properties of Thermoplastic Polymers

Optical Properties

There are a number of optical properties of importance with thermoplastic polymers including clarity, haze, color, transmittance, reflectance, gloss and refractive index.

The clarity of a film indicates the degree of distortion of an object when viewed through the film, "see-through" clarity referring to the ability of the film to resolve fine details of fairly distant images viewed through the film. It should not be confused with haze as hazy materials often have quite good see-through clarity. In order to achieve a product with a high clarity, it is important that the refractive index is constant throughout the sample in the line of direction between the object in view and the eye. The presence of interfaces between regions of different refractive index will cause scattering of the light rays.

Amorphous polymers free from fillers or other impurities are transparent unless chemical groups which absorb visible light radiation are present. Generally, crystalline polymers are translucent, but they will be transparent when the crystal structures such as spherulites are smaller than the wavelength of light. However, if the structures are greater in diameter than the wavelength of light, the light waves will be scattered if the crystal structures have a different refractive index to that of the amorphous regions. The clarity of crystalline plastics can be improved by quenching or by random copolymerisation.

Since refractive index is dependent on density, it follows that where the crystalline and amorphous densities of polymers differ, there will be a difference in refractive index. Thus, thick polyethylene objects will be opaque since they cannot be quenched rapidly, and the spherulites formed have a significantly higher density (about 1010 kg m^{-3}) than the amorphous region (840–850). In the case of polypropylene, the difference is less marked (crystal density 940; amorphous density 850) and moldings are more translucent. Generally, the refractive index value for plastics is around 1.5.

Haze can be taken as a measure of the "milkiness" or "cloudiness" of an otherwise transparent polymer, and is often the result of surface imperfections, particularly in the case of thin films. The appearance of haze with consequent loss of contrast is caused by light being scattered by the surface imperfections, or by in homogeneities in the film due to voids, large crystallites, incompletely dissolved additives or cross-linked material. Internal haze does not arise with amorphous polymers; with crystalline polymers it increases with the degree of crystallinity and size of spherulites or other forms of crystal aggregates, as well as with the ratio of the density between the crystalline phase and the amorphous-phase.

Methods for measuring haze are specified in ASTM D 1003 and in BS 2782. Haze is defined in these standards as the percentage of transmitted light which, in passing through a specimen, deviates by more than 2.5° on average from an incident parallel beam by forward scattering from both surfaces and from within the specimen.

When light falls on a material, some is transmitted, some is reflected and some absorbed. The transmittance is the ratio of the light passing through to the light incident on the material, and the reflectance the ratio of the reflected light to the incident light.

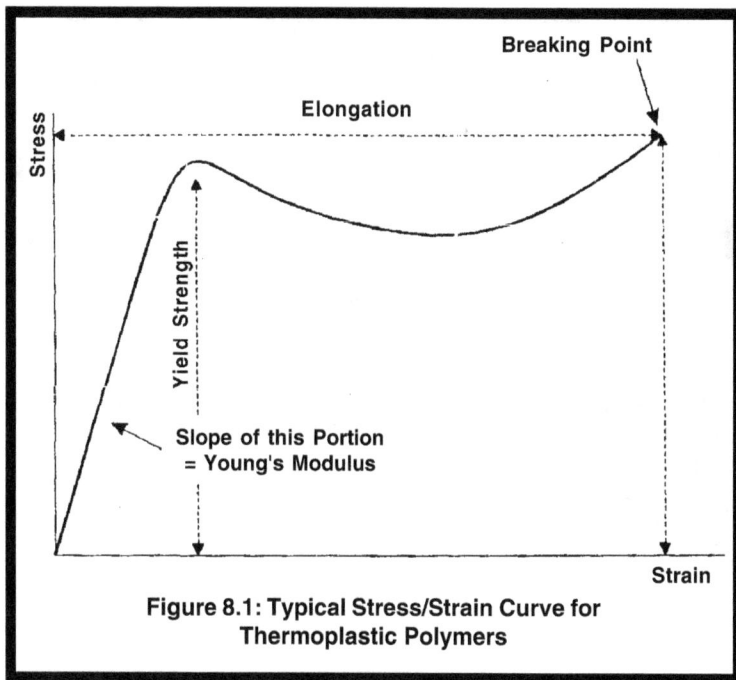

Figure 8.1: Typical Stress/Strain Curve for Thermoplastic Polymers

The gloss (strictly speaking specular gloss) of a film has been defined as the degree to which a surface simulates a perfect mirror in its capacity to reflect incident light. Thus, it is a function of the reflectance and the surface finish of a material. Where transmittance and reflectance do not add up to unity, then some of the light waves are absorbed; if this absorption does not occur uniformly over the visible spectrum, the material appears colored. Gloss is determined by standard tests (ASTM D 523, D 1834 and D 2457, and BS 2782) using a device (known as a glossmeter) which measures the percentage of the light, incident at an angle (usually 45°) to the surface of the film, that is reflected at the same angle. The fraction of the original light that is reflected is the gloss of the sample.

Impact Strength

The impact properties of polymeric materials are directly related to the overall toughness of the material, toughness being the ability of the polymer to absorb applied energy. The area under the stress-strain curve is directly proportional to the toughness of the material.

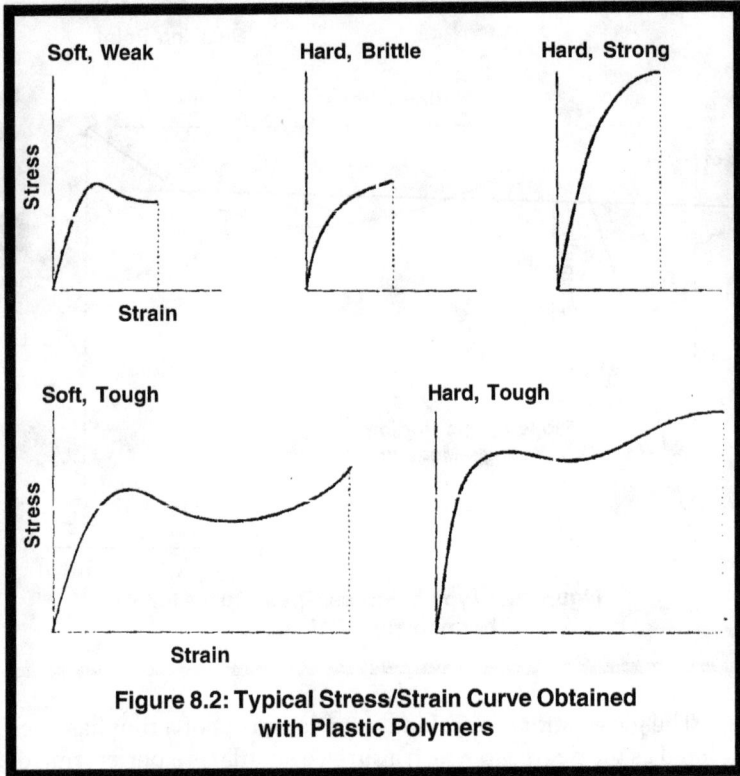

**Figure 8.2: Typical Stress/Strain Curve Obtained
with Plastic Polymers**

The impact strength of a film is a measure of its ability to withstand shock loading.

The pendulum impact tester, also known as the Izod Test, can be used to measure impact strength. An impacting head on the end of a pendulum is swung through an arc into and through the sample. This test measures the difference between the potential energy of the pendulum at the maximum height of its free swing, and the potential energy of the pendulum after rupture of the sample. This difference in energy is defined as impact strength and is useful in predicting the resistance of a material to breakage from dropping or other quick blows.

Another method (ASTM D 2176 Method B) utilizes the M.I. T. Fold Endurance tester. It has an upper spring-loaded clamp which moves only up and down, an oscillating folding head which supports the smooth, folding surfaces, a driven device to provide rotary oscillating motion to the folding head, and a counter to register the number of folds. The sample is clamped into the folding head and the upper spring-loaded clamp depressed by a weight equivalent to the desired tension on the sample. The sample is then clamped into the upper clamp and the tension applied. The motor is started and the sample folded until it is severed at the crease.

The above two tests may not always be applicable to polymeric films; the number of folds required to break the test specimen depends on its thickness. Thus nylon film 2.54 μm thick may withstand over 200,000 double folds in the M.I.T. tester before failing, but 25.4 μm thick nylon may fail after fewer than 100,000 double folds.

8.2 Tensile Properties

The four properties tensile and yield strength, elongation and Young's modulus are considered under a single heading because the same equipment is used for measuring each of them. The basic principles of tensile testing (tensile meaning capable of being drawn out or stretched) can be found in any introductory physics textbook; a discussion relating to plastics is given by Shah. The tensile test is the simplest mechanical test to visualize; in it an increasing tensile stress is applied to a material and the resulting changes in length are monitored.

The *stress*, σ, is defined as the force per unit cross-sectional area of material. Thus, if the force is F Newtons and the cross-sectional area is A square meters, then the stress is given by:

$$\sigma = F/A \ N\,m^{-2}$$

The *strain*, ε, is defined as the fractional change in length of the material, *i.e.*, $\Delta l/l_o$, where l_o is the initial length and Δl the change in length. It is expressed as a dimensionless ratio.

Tensile strength (or more accurately, ultimate tensile strength) is the maximum tensile strength which a material can sustain and is taken to be the maximum load exerted on the test specimen during the test, divided by the original cross-section of the specimen.

Yield strength is the tensile stress at which the first sign of a non-elastic deformation occurs and is the load at this point (known as the yield point) divided by the original cross-section of the specimen.

Elongation is usually measured at the point where the film breaks and is expressed as the percentage of change of the original length of the material between the grips of the testing machine. Its importance is as a measure of the film's ability to stretch, a large value for elongation indicating that the material will absorb a large amount of energy before breaking.

Young's modulus, or the modulus of elasticity, is the ratio of stress to strain over the range for which this ratio is constant, *i.e.*, up to the yield point. It is a measure of the force that is required to deform the film by a given amount and so it is also a measure of the intrinsic stiffness of the film.

Yield strength is usually more important than ultimate tensile strength, especially during the passage of the film through packaging or printing equipment where a sudden "snatch" could cause a non-reversible distortion and thus, for example, out-of-register printing.

Both yield point and elongation are important properties during the unwinding of plastic films. There is the danger of uneven stretching of the film if the elongation is high unless special handling techniques are used. Too low an elongation should also be avoided as any sudden unbalance in the unwind operation could lead to breaking of the film. Because a certain amount of tension is necessary during unwind, the possibility exists that films with a low yield strength could be stressed beyond their yield point.

One problem which arises when considering the strength of thermoplastic polymers is the time and temperature dependence of the changes which occur under mechanical stress, particularly in the case of thermoplastics. This time-dependence is well illustrated by the behavior of thermoplastics under tensile stress. This is measured by gripping the test material between fixed and moving clamps which are capable of separating at a range of speeds, the stress in the sample being measured continuously and recorded on a chart. At very slow speeds the molecules can readily disentangle, and the measured tensile strength depends largely on the magnitude of the weakest intermolecular forces. At faster speeds, on the other hand, there is little time for either disentanglement or slipping and the breaking point will not happen until the largest intermolecular forces have been overcome.

Methods for testing tensile properties of thermoplastic films are described in ASTM D 882 and BS 2782-301. It is usual when carrying out a tensile strength determination to plot the stress against the strain, *i.e.,* the load against the elongation. Such a stress/strain curve can give a great deal of information. For instance, brittle plastics will break at a much earlier stage of the curve, sometimes before the yield point has been reached.

A great deal of information about the material can be obtained from the shape of its stress/strain curve. In addition to the numerical values for tensile strength, Young's modulus, elongation, etc. it is possible to obtain some idea of the toughness of the material by measuring the area under the curve. This area is a measure of the energy needed to break the test specimen and hence is directly related to toughness.

8.3 The Transfer of Substances in Polymeric Materials

The solution and transport behaviour of low molecular weight substances in polymeric materials has become increasingly important in recent years, with the widespread use of polymer films and rigid plastics for food packaging.

Unfortunately, there are many examples of foods packaged with an apparent lack of proper consideration of the effects of the end use environment on properties, or of limitations imposed on performance due to unfavorable solution or transport characteristics. The plasticization of polymers by sorption of ambient vapors or liquids resulting in the decrease in mechanical properties and the loss of beverage components (*e.g.,* CO_2, flavor, etc.) are just two of many examples which could be cited. An objective of research in this field is to establish mechanisms and expressions relating solubility and transport with the molecular properties and characteristics of the components.

The protection of foods from gas and vapor exchange with the environment depends on the integrity of packages (including their seals and closures), and on the permeability of the packaging materials themselves. There are two processes by which gases and vapors may pass through polymeric materials:

1. A pore effect, in which the gases and vapors flow through microscopic pores, pinholes, and cracks in the materials; and

2. A solubility-diffusion effect, in which the gases and vapors dissolve in the polymer at one surface, diffuse through the

polymer by virtue of a concentration gradient, and evaporate at the other surface of the polymer. This "solution-diffusion" process (also known as "activated diffusion") is described as true permeability.

Most polymers when sufficiently thin exhibit both forms of permeability. Porosity falls very sharply as the thickness of a polymer is increased, reaching virtually zero with many of the thicker types of commercially available materials. True permeability, however, varies inversely as the thickness of the material and hence, cannot be effectively eliminated merely by increasing the material thickness.

This chapter is concerned with aspects of the solution, diffusion and permeation of gases and vapors ("permeants") in effectively non-porous polymeric materials. Mention will not be made of systems involving liquid solvents or higher concentrations of sorbed solvent vapors. Such systems often show behavior which seemingly deviates strongly from that observed with gases or low concentrations of sorbed vapors. Such deviations are classified as anomalous diffusion, Case II, or Super Case II behavior and the interested reader is referred to a more detailed review.

The first recorded observation of the permeation of gas through a membrane appears to be that of Thomas Graham who in 1829 observed the permeation of CO_2 into a wet pig's bladder. In 1831 J.K. Mitchell, an American physician and the inventor of the toy rubber balloon, discovered that his balloons collapsed at different rates when they were filled with different gases.

The next major step came in 1855 when Adolf Fick proposed his law of mass diffusion by analogy with Fourier's law for heat conduction and Ohm's law for electrical conduction. In 1866 Graham postulated that the permeation process entailed dissolution of the penetrant, followed by transmission of the dissolved species through the membrane as though through a liquid, a process Graham called "colloidal diffusion". His concept is similar to that which is still used today, known as the solution-diffusion model. The subject was placed on a quantitative basis in 1879 by Von Wroblewski who showed that the solubility of gases in rubber obeyed Henry's law, and combined this with Fick's law to obtain the now familiar expression relating permeation rate and the area and thickness of the membrane.

Under steady state conditions, a gas or vapor will diffuse through a polymer at a constant rate if a constant pressure difference is maintained across the polymer. The diffusive flux, of a permeant in a polymer can be defined as the amount passing through a plane (surface) of unit area normal to the direction of flow during unit time, *i.e.*:

$$J = Q/A \cdot t \qquad \text{......(8.1)}$$

where Q is the total amount of permeant which has passed through area A during time t.

The relationship between the rate of permeation and the concentration gradient is one of direct proportionality and is embodied in Fick's first law:

$$J = -D \frac{\delta C}{\delta x} \qquad \text{......(8.2)}$$

where J is the flux (or rate of transport) per unit area of permeant through the polymer, c is the concentration of the permeant, D is defined as the diffusion coefficient and $\delta c/\delta x$ is the concentration gradient of the permeant across a thickness δx.

Consider a polymeric material X mm thick, of area A, exposed to a permeant at pressure p_1 on one side and at a lower pressure p_2 on the other. The concentration of permeant in the first layer of the polymer is c_1 and in the last layer c_2.

If x and $(x + \delta x)$ represent two planes through the *polymer* at distances x and $(x + \delta x)$ from the high-pressure surface, and if the rate of permeation at x is J mL per sec, and at $(x + \delta x)$ is $I + (\delta J/\delta x)\delta x$, then the amount retained per unit volume of the polymer is $(\delta J/\delta x)$. This is equal to the rate of change of concentration with time:

$$\frac{\delta}{\delta x}(J) = -\frac{\delta c}{\delta t} \qquad \text{......(8.3)}$$

A negative sign is used because the concentration of permeant decreases across the material. Under steady state conditions, $\delta c/\delta t = O$ and J = constant. When the concentration gradient is zero (*i.e.*, $c_1 = c_2$) there will be no permeation.

If equation 8.2 is substituted into equation 8.3 then:

$$\frac{\delta}{\delta x}(J) = \frac{\delta}{\delta x} - D\frac{\delta c}{\delta x} = -\frac{\delta c}{\delta t} \qquad ....(8.4)$$

and on rearrangement of the terms:

$$\frac{\delta c}{\delta t} = \frac{\delta}{\delta x} D\frac{\delta c}{\delta x} \qquad ....(8.5)$$

and

$$\frac{\delta c}{\delta t} = D\frac{\delta^2 c}{\delta x^2} \qquad ....(8.6)$$

Equation 8.6 is a simplified form of Fick's second law of diffusion and applies under circumstances where diffusion is limited to the x-direction and D is independent of concentration. There are extensive discussions of mathematical methods for solving the diffusion equation, together with solutions for many of the more common situations.

8.4 Permeant Relationships

For monocondesable gases, it has been predicted that the permeability ratio of a pair of gases will be relatively constant over a series of polymers.

It can also be seen that regardless of the film material, O_2 permeates about four times as fast as N_2, and CO_2 permeates about six times as fast as O_2 and 24 times as fast as N_2. There is no pattern for the ratios of water vapor to any of the three gases.

It may be thought to be strange that CO_2, the largest of the three gas molecules, has the highest permeability coefficient. In fact, it has the lowest diffusion coefficient of the three gases as would be expected from its relative size, but its permeability coefficient is the highest because its solubility coefficient S is much greater than that for the other gases.

Table 8.1: Diffusion, Solubility and Permeability Coefficients for Low Density Polyethylene Film at 25°C to CO_2, O_2 and N_2

	$D \times 10^6$ $cm^2\ s^{-1}$	$S \times 10^2$ $mL(STP)\ mL^{-1}\ atm^{-1}$	$P \times 10^{10}$ $[mL(STP)\ cm$ $cm^{-2}\ s^{-1}\ (cm\ Hg^{-1})]$
CO_2	0.37	25.8	12.6
O_2	0.46	4.78	2.88
N_2	0.32	2.31	0.969

It has been argued that the permeability coefficient for a particular permeant/polymer system is a product of a factor F determined by the nature of the polymer, a factor G determined by the nature of the permeant, and an interaction function H. The latter is taken as unity where little or no interaction occurs, and becomes larger the greater the degree of interaction between the permeant and the polymer.

Thus, the permeability coefficient of a polymer x to a permeant k can be expressed as:

$$P_{xk} = F_x\,G_k\,H_{xk} \qquad\qquad ......(8.7)$$

If H_{xk} is taken as unity, the ratio of the permeability coefficients of a polymer x to permeants k and j can be shown to be the same as the ratio between their respective G factors:

$$\square \qquad\qquad \frac{P_{xk}}{P_{xj}} = \frac{G_k}{G_j} \qquad\qquad ......(8.8)$$

The above ratio would be almost independent of the nature of the permeant.

Similarly for the permeability coefficients of two polymers x and y and a permeant k:

$$\square \qquad\qquad \frac{P_{xk}}{P_{yk}} = \frac{F_x}{F_y} \qquad\qquad ......(8.9)$$

The above ratio would be independent of the nature of the polymer material.

From a knowledge of various values of P it is possible to calculate F values for specific polymers and G values for specific gases if the G value for one of the gases (usually nitrogen) is taken as unity. For non-condensable gases it has been predicted that the ratios of the permeability coefficients of a pair of gases will be relatively constant over a series of polymers.

Chapter 9

Important Plastics Processing Methods

9.1 Extrusion

Extrusion is one of the most important plastics processing methods in use today. Most plastic materials are processed in extruders and commonly pass through two or more extruders on their way from the chemical reactor to the finished product. All of the thermoplastics are formed into sheet or film by the process of screw extrusion.

The heart of the extruder is the Archimedean screw which revolves within a close-fitting, heated barrel. It is capable of pumping a material under a set of operating conditions at a specific rate, depending on the resistance at the delivery end against which the extruder is required to pump. The extruder resembles a mincer into which granules are fed, heated and compressed until they fuse into a melt which is forced through a slot or circular die. Sufficient heat is produced by the energy of the screw to allow the process to continue with very little heat input.

The standard single screw extruder (Figure 9.1) receives cold powder or granules through a hopper to a throat at one end of the

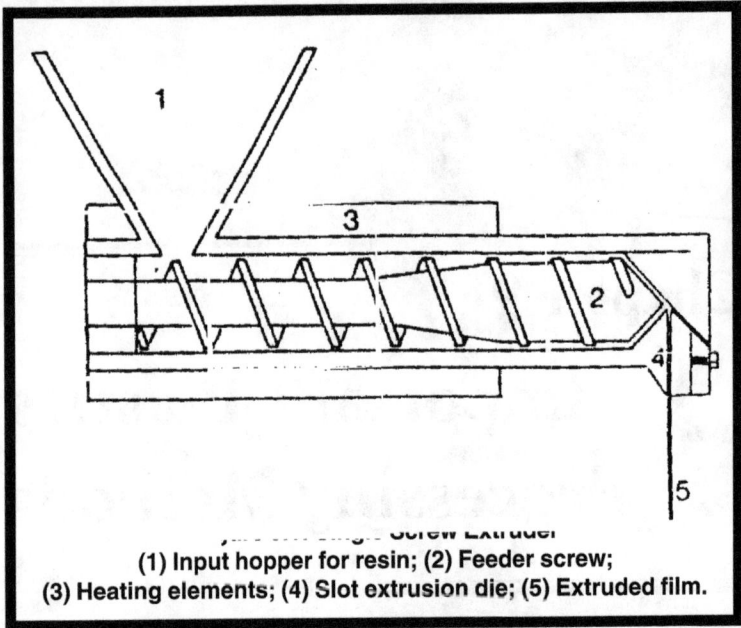

(1) Input hopper for resin; (2) Feeder screw;
(3) Heating elements; (4) Slot extrusion die; (5) Extruded film.

barrel and delivers it to the compression zone of the screw. In this section the diminishing depth of thread causes a volume compression and an increase in the shearing action of the material. It melts and is converted into a homogeneous mass by contact with the heated walls of the barrel and by the heat generated by friction. Generally external heating is only required at the start of the run, the frictional or exothermal heating being sufficient for steady-state operation. After the compression zone, the melt passes through the metering zone where the flow is stabilized, before being pumped through the die which determines its final form. The output from the die is known as the extrudate.

The screw is the most important component of the extruder and different designs are used for extruding different polymers. Extruder screws are characterized by their length to diameter ratios (commonly abbreviated to L/D ratios) and their compression ratios, this being the ratio of the volume of one flight of the screw at the infeed end to the volume of one flight at the die end. L/D ratios commonly used for single screw extruders are between about 15 : 1 to 30 : 1, while compression ratios can vary from 2 : 1 to 4 : 1.

There are basically two processes by which the extruded thermoplastic can be converted into film: the tubular process and the flat film process.

In the tubular (or blown film) process, a thin tube is extruded (usually in a vertically-upward direction), and by blowing air through the die head, the tube is inflated into a thin bubble. This is cooled, flattened out and wound up. The ratio of bubble diameter to die diameter is known as the blow-up ratio. Most low density polyethylene blown films used in packaging are made using blow-up ratios of between 2.0 and 2.5 : 1. Blown film extrusion can produce defects such as variations in film thickness, surface defects, low tensile and impact strength, haze, blocking and wrinkling.

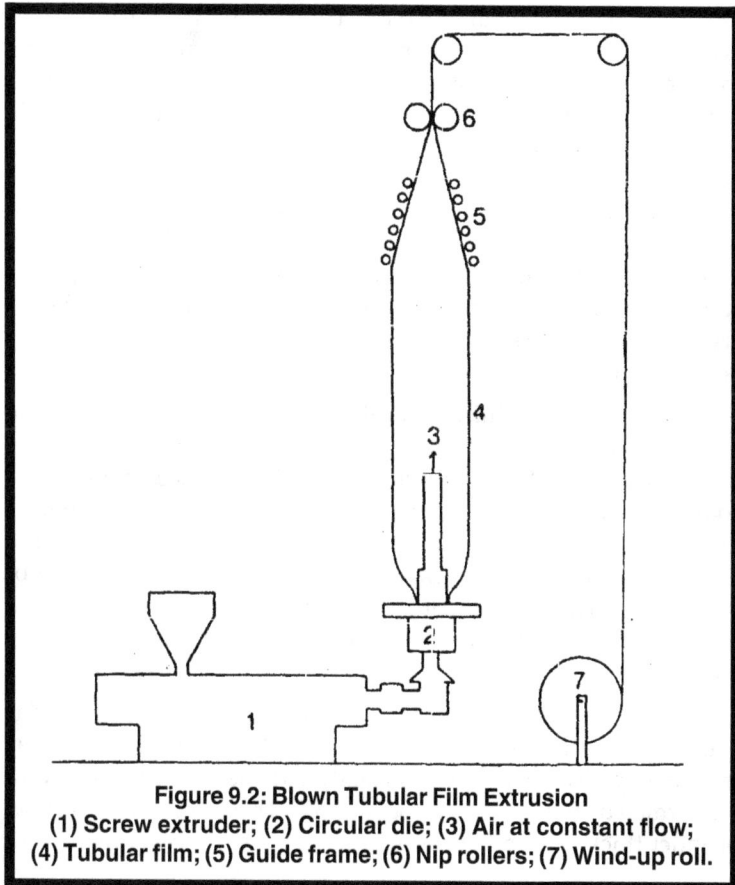

Figure 9.2: Blown Tubular Film Extrusion
(1) Screw extruder; (2) Circular die; (3) Air at constant flow;
(4) Tubular film; (5) Guide frame; (6) Nip rollers; (7) Wind-up roll.

The properties of the film depend strongly on the polymer used and the processing conditions. The higher the density, the lower the flexibility and the greater the brittleness. The higher the molecular weight, the greater the tensile strength and resistance to film brittleness at low temperatures, but the lower the transparency.

In flat film (also known as cast film or slit die) extrusion, the molten polymer is extruded through a slit-die into a quenching water bath or onto a chilled roller. In both cases, rapid cooling of the extruded film is most important. The ratio of the haul-off rate to the natural extrusion rate is referred to as the draw-down ratio. Draw-down ratios between 20 : 1 and 40 : 1 are typical.

A comparison of the tubular and flat film processes lists among the advantages of tubular film the fact that the mechanical properties are generally better, and that the process is easier and more flexible to operate. The cost for making wide tubular film is much lower than for wide cast film due to the cost of precision grinding long chill rolls. The advantages of the flat film process include less thickness variation, very high outputs and superior optical properties. This latter advantage is a consequence of the quicker cooling which can be achieved in the flat film process where cooling is by conduction, compared with the tubular film process, where cooling is by convection. Slower cooling permits the formation of more and larger crystals in the film, leading to haze which arises from light scatter between the crystal interfaces.

By using combining adaptors, it is possible to extrude simultaneously two or more different polymers which fuse at the point of film formation into a single web. Such a process is known as coextrusion and permits the production of a single web having, for example, barrier properties not possessed by anyone of the component polymers. A two-component slot die is capable of producing a two- or three-layer film from two materials, while a three-component die can produce a five-layer film from three materials.

Film thicker than 0.25 mm is normally defined as sheet, and is commonly thermoformed into objects that hold their shape such as trays or cups. The polymer is normally extruded horizontally into a nip formed by two hardened cooling rolls. These determine the final product thickness and surface finish. Sheets up to about 1.3 mm

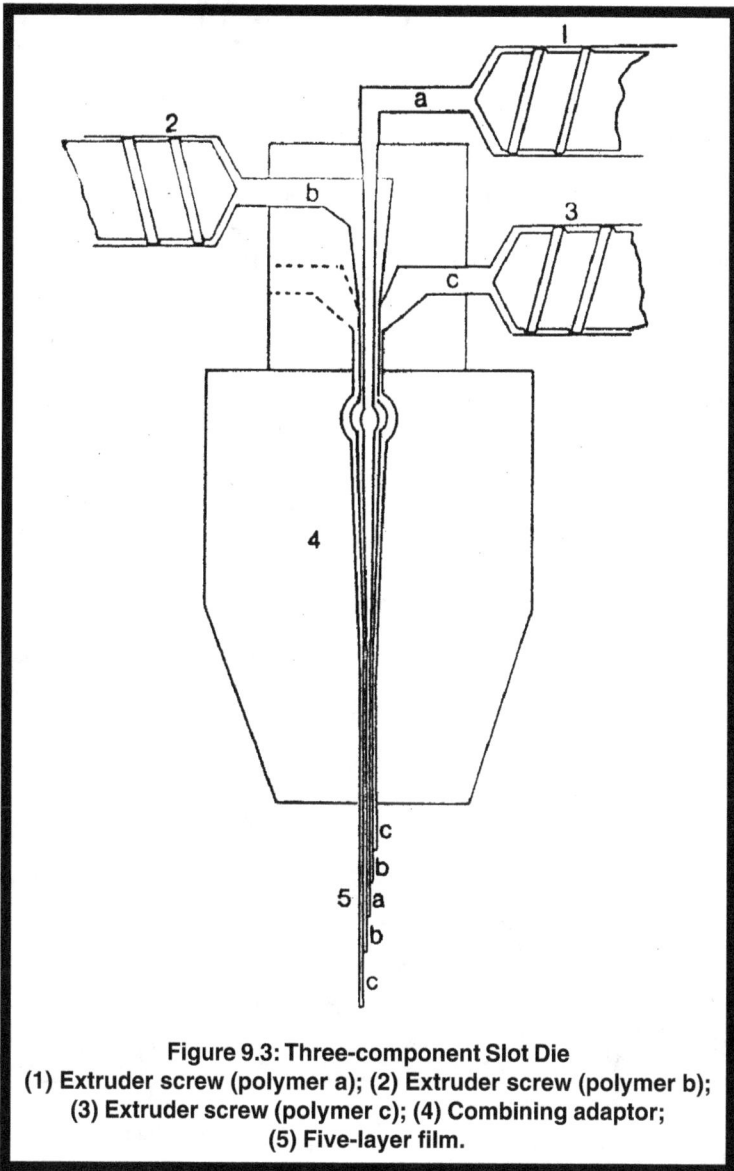

Figure 9.3: Three-component Slot Die
(1) Extruder screw (polymer a); (2) Extruder screw (polymer b);
(3) Extruder screw (polymer c); (4) Combining adaptor;
(5) Five-layer film.

thick can be wound onto rolls, thicker sheet being cut to desired lengths.

9.2 Extrusion Blow Molding and Injection Blow Molding

Extrusion blow molding uses many arrangements for making and forming the parison, and the reader is referred to Irwin for a detailed discussion. In the simplest method, a mold is mounted under the die and the parison extruded between the open halves of the mold. When the parison reaches the proper length, the extruder is stopped and the mold closes around the parison.

A blow pin mounted inside the die head allows air to enter and blow the parison into the final container shape. The shape of the bottle or jar is defined, but the distribution of material (and thus wall thickness) is less well controlled. The cycle restarts after the part has cooled and the mold opened. To use the full capacity of the extruder, numerous systems are used.

One development is to use more than one mold, moving the filled mold away to cool while another is moved into position to receive the next section of extruded tube. The molds can be reciprocating ones, or can be mounted on a rotary table. In several

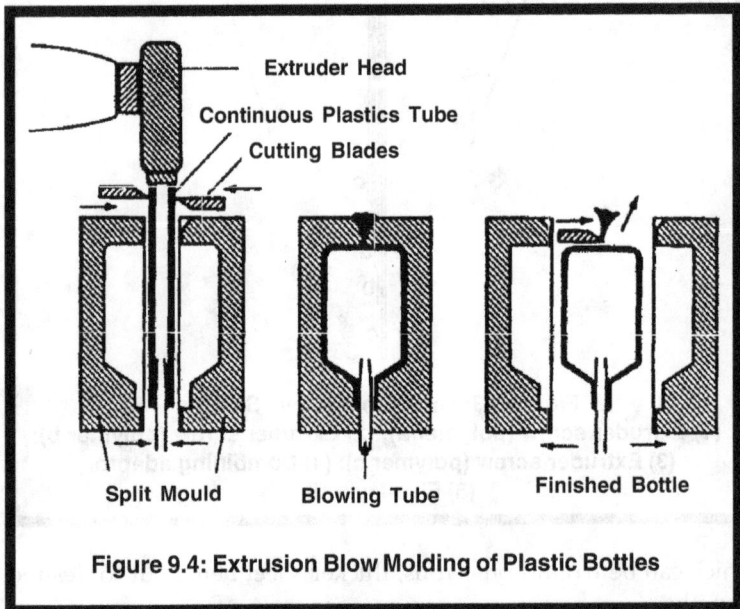

Figure 9.4: Extrusion Blow Molding of Plastic Bottles

food packaging applications such as fresh milk, the bottles are blow molded and filled on line in a continuous operation.

Extrusion blow molding is widely used with the following resins: high density polyethylene, polypropylene, poly(vinyl chloride) and acrylonitrile copolymers. The common grades of poly(ethylene terephthalate) cannot be extrusion blown.

A related development has been the production of co-extruded bottles, where two or more extruders, each handling a different plastics material, produce a multi-layer parison having the desired properties. For example, a high-barrier, high cost material might be sandwiched between layers of a relatively low cost material to give a bottle with the desired barrier properties at an economical price.

Injection blow molding is a non-continuous cyclic process most closely resembles the blowing of glass bottles. The parison is formed in one mold and then, while still molten, is transferred to a second mold where blowing with compressed air forms the final shape. After cooling, the mold is opened and the bottle ejected. Several molds must be available if the injection molding machine is to operate near full capacity. The major advantage of injection molding over extrusion blow molding is that the process is virtually scrap-free, the finished parts usually requiring no further trimming, reaming or other finishing steps. In addition, the dimensions of the bottle (including the neck) show very little variation from bottle to bottle, and with some materials, improved strength and clarity are obtained due to the effect of a limited degree of biaxial orientation.

The resins most commonly used for injection blow molding are high density polyethylene, polypropylene, polystyrene and poly(vinyl chloride). Recently, poly(ethylene terephthalate) has been injection blow molded, and is likely to replace poly(vinyl chloride) in some applications.

In recent years, co-injection blow molding has been developed using two or three injection units working with one mold to produce either a part component or a perform which is later blow-formed using compressed air inside a mold to make a bottle or jar. The various component materials are metered into cavities in such an order that the barrier material flows through the main structural material to create a multi-layer structure. This process is used to produce five

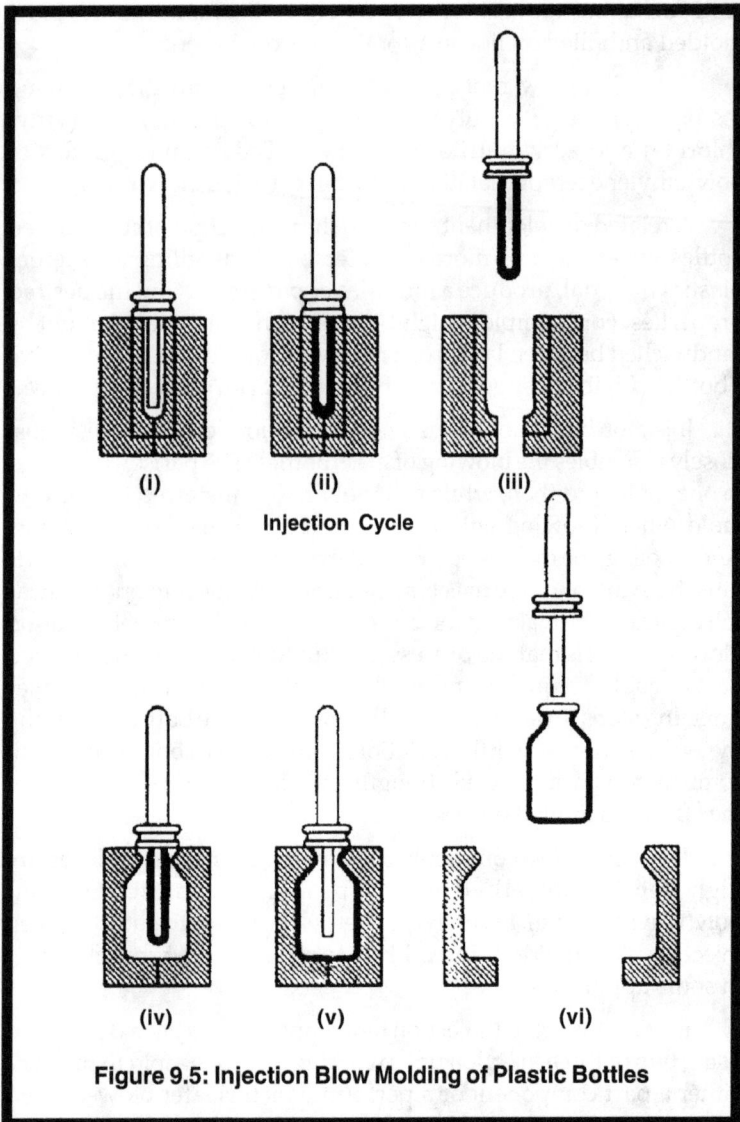

(i) (ii) (iii)

Injection Cycle

(iv) (v) (vi)

Figure 9.5: Injection Blow Molding of Plastic Bottles

layer retortable containers from three materials, typically polypropylene as a structural layer and ethylene vinyl alcohol copolymer as a barrier layer with tie layers in between.

An alternative process for the production of retortable plastic containers has been developed by Hitek in Australia. The process is known as Cuspation-Dilation and was developed from the observation that a sheet of hot plastic could not be perforated by a metal needle; instead the sheet formed a cusp. Basically the process involves a set of metal fingers inside a female mold which stretch the molten plastic sheet into a series of cusps. The fingers then dilate to further stretch the material into the shape of the mold. Containers with very even cross section can be made using this process, without the thickness that would be required in an injection molded product.

9.3 The Orientation of Polymer Films

Orientation of polymer films is a means of improving their strength and durability in order to broaden their scope of application and make them serviceable in thinner gauges. Films may be oriented in either one direction (uniaxial orientation) or, more commonly, in two directions, usually at right angles to each other (biaxial orientation). Virtually all thermoplastics can be oriented to some extent, but amorphous films can be more readily oriented than crystalline films.

The first biaxial orientation process was developed in Germany about 1935, and during World War II oriented polystyrene film was made in Europe for capacitors and coaxial cable insulation. In 1948 Dow Chemical Company developed the Saran (vinylidene chloride/ vinyl chloride copolymer) resins and Dewey and Almy (now the Cryovac Division of W.R. Grace) developed the film process and introduced Saran shrink film to the market. In 1952 E.I. du Pont de Nemours & Co., Inc., began selling oriented poly(ethylene terephthalate) film.

In essence orientation of thermoplastic film (whether uniaxial or biaxial) is a process of stretching the material in such a manner as to line up the molecular chains in a predetermined direction. Once lined up, the ordered arrangement is frozen in the strained condition. Biaxially oriented films possess superior tensile and impact strengths, improved flexibility, clarity, stiffness and toughness, and increased shrinkability. Gas and water vapor permeability may also be reduced, generally by 10 to 50 per cent depending on the type of polymer, and the degree and temperature of orientation.

Gas and water vapor permeabilities for amorphous polymers (*e.g.*, polystyrene and polyesters) appear to be nearly identical for both oriented and unoriented films. Crystalline polymers (*e.g.*, polypropylene and vinylidene chloride/vinyl chloride copolymer) show significant reduction in water vapor peremeability when oriented. This difference is greatest at low degrees of crystallinity (10–15 per cent) and gradually becomes less as the degree of crystallinity increases, until at 40–50 per cent crystallinity, no differences are discernible. The gas permeability rates are largely dependent on the amporphous. The gas permeability rates are largely dependent on the amorphous content, which outweighs any effect introduced by orientation.

On the other hand, orientation generally has a detrimental effect on elongation, ease of tear propagation and the sealability of the film. The heat sealability range is narrowed and the film may vary in properties with age. Oriented film cannot be easily heat sealed because it shrinks and puckers at temperatures below the sealing temperature. Although various mechanical devices ('point sealing') have been designed to mitigate the shrinkage, a more suitable solution is to apply a surface coating of some thermoplastic having a lower melting point. For example, oriented polypropylene (OPP) may be coated with a dispersion of a copolymer of vinylidene chloride and vinyl chloride, or a copolymer of polypropylene with a small quantity of ethylene.

Among the more common commercially oriented films are poly(ethylene terephthalate), polyamide, vinylidene chloride/vinyl chloride copolymers, polypropylene and low density polyethylene, the latter commonly being irradiated before blowing into film. Because radiation cross-links the molecules, the film can be stretched without becoming fluid at the melting point of a non-irradiated film, resulting in greatly improved tensile strength and shrink tension compared to non-irradiated low density polyethylene. High density polyethylene is not oriented because its very rapid crystallization limits the extent to which it can be stretched. When the resin is blended 70 : 30 with low density polyethylene, the rate of crystallization is slowed. The crystallinity imparts properties similar to those achieved through radiation cross-linking. The largest application of orientation techniques is in the manufacture of OPP which results in a considerable improvement in its barrier properties.

In the case of crystalline polymers, the action of orientation induces additional crystallization, with the crystalline structure

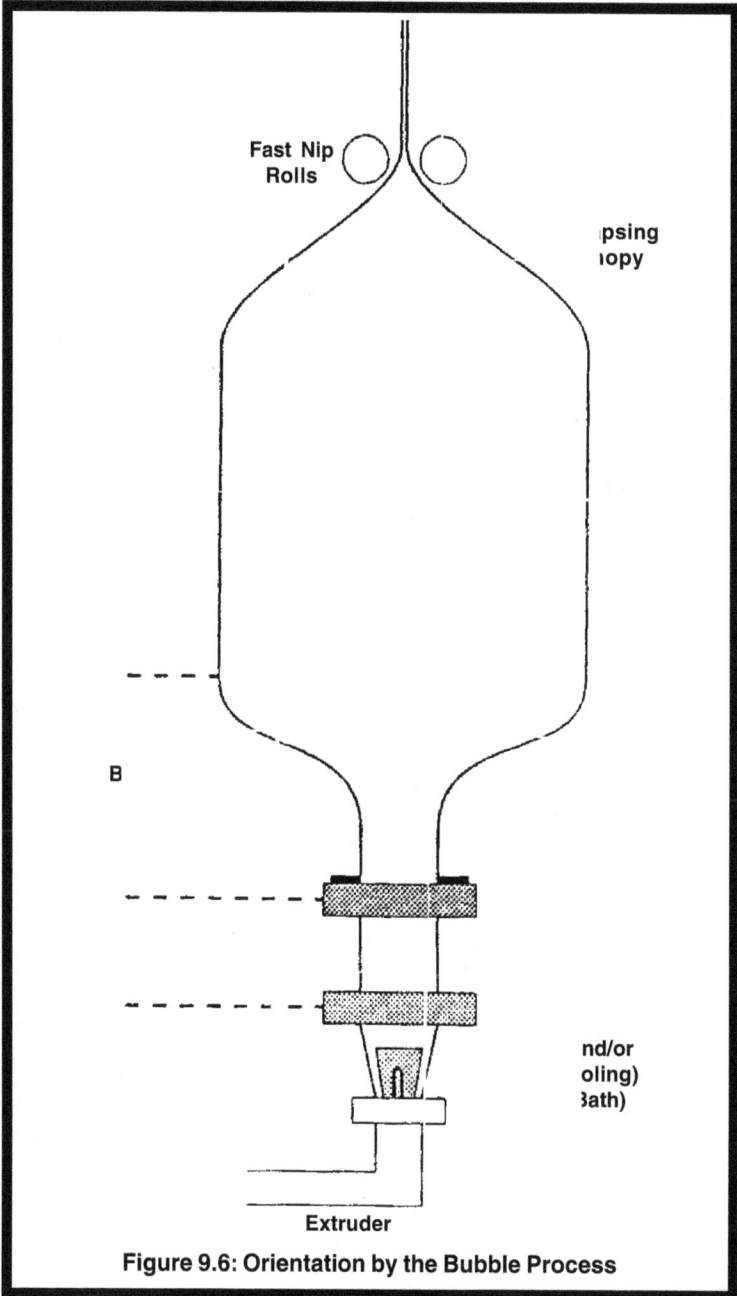

Fast Nip Rolls

psing
opy

B

nd/or
oling)
Bath)

Extruder

Figure 9.6: Orientation by the Bubble Process

aligned in the direction of stretching. The induced crystallinity is general and does not occur in spherulite form; therefore, oriented films usually have a high degree of clarity, because of the relative absence of spherulites which cause light scattering.

For many applications, shrinkage is not desirable and a greater degree of heat stability is required. Films can be annealed by application of heat to partially relax the forces while maintaining the film in a highly stretched condition. It is then cooled to room temperature and the restraint on the film released. Such a film is referred to as heat set and will not shrink if heated to below the annealing temperature. The procedure of annealing does result in some reduction in dimension in the stretched direction or directions.

The most common method used to orientate a thermoplastic film is to stretch it after it has been heated to a temperature at which it is soft. This temperature is below the flow temperature at which the molecules would glide readily past one another when the material is stressed, but above the glass transition temperature. As a result of this stretching, the direction of the molecules changes towards that in which the material is stressed, and the molecules are extended like springs. The temperature is then dropped below the softening point of the material while the molecules are held in this configuration, so that the molecules are frozen in the strained position.

Films can be oriented using two processes–flat sheet and tubular. In the flat sheet or tentering process, thick (500–600 µm) cast film is fed to a system of differential draw rolls which are heated to bring the film to a suitable temperature below its melting point. The film is stretched in the machine direction and then fed to a tenter frame where a series of clips (mounted side by side on endless chains) grasp both edges of the film and draw it transversely as it travels forward at an increasing speed. Draw ratios in both directions normally vary between 4 : 1 and 10 : 1. After tentering, the film is passed over a cooling roller and reeled up. The two operations can also be carried out sequentially, *i.e.,* the film is stretched in the machine direction first, and then fed into a tenter frame where it is drawn transversely.

In the tubular or bubble process, molten polymer is extruded from an annular die and then quenched to form a tube. The tube is

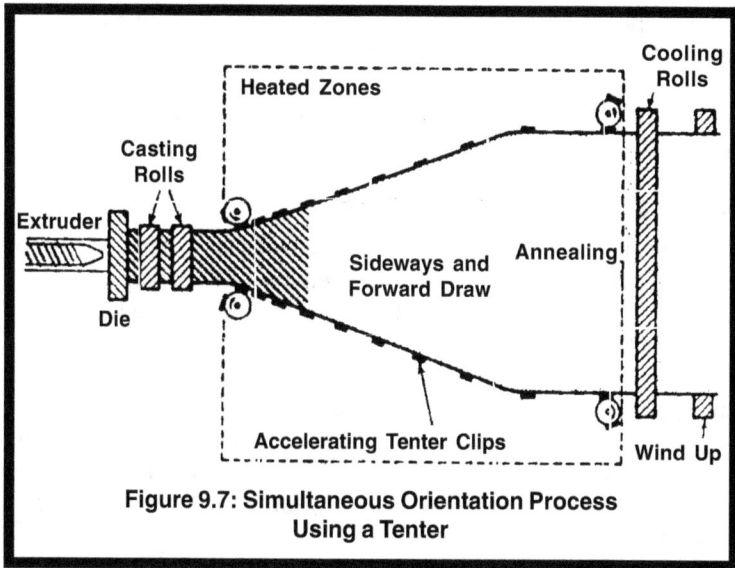

Figure 9.7: Simultaneous Orientation Process Using a Tenter

flattened by passing through nip rolls and reheated to a uniform temperature. The air pressure in the tube is increased to expand the film transversely, the draw ratio being varied by adjusting the volume of entrapped air. Pinch or collapsing rolls at the end of the bubble are run at a faster speed than rolls at the beginning of the bubble, thus causing drawing of the film in the machine direction. The film is then wound-up.

The amount of orientation imparted to a film depends on the stretching temperature, the amount of stretching, the rate of stretching and the quench. Quenching is carried out either by extruding the web onto a chill roll or by passing it through a quench tank prior to orientation. Generally orientation is increased by decreasing the stretching temperature, increasing the amount of stretch, increasing the rate of stretch, and increasing the amount of quench. Films such as vinylidene chloride/vinyl chloride copolymer and polypropylene which have glass transition temperatures below room temperature show an appreciable crystallization rate even at room temperature, and therefore, have to be quenched and oriented immediately after extruding.

The potential energy stored in the extended molecules is the so-called "elastic memory" characteristic of oriented, non-heat set thermoplastics. When such a film is reheated to its orientation temperature, it shrinks as the molecules tend to return to their original size and spatial arrangement.

9.4 The Shrink and Stretch Films

The first shrink film was Saran and it was used initially for frozen poultry. The unfrozen birds were placed in a bag, a vacuum was applied to remove air and the bag was then sealed, typically with a metal clip. On immersion in hot water, the bag shrank tightly around the bird which was then frozen, commonly by passage through a glycol bath at −30°C. The close contour wrap ensured rapid heat transfer during freezing since there were no air pockets inside the package. This latter fact, coupled with the impermeability of the film, prevented freezer burn, *i.e.*, sublimation of water vapor from the surface of the frozen poultry.

Today a range of shrink films made from a variety of polymers is available. With the exception of Saran, most of these other shrink films require temperatures above 100°C to obtain a suitable degree of shrinkage, necessitating the development of hot air tunnels or heat guns.

Three properties of shrink films are important when selecting a film for a particular application. The first is the range of temperature over which a film will shrink. The lower the shrink temperature, the simpler and less expensive the shrink process. Films with a wide softening range are usually preferable since this makes temperature control of the heating equipment less critical.

The degree of shrinkage is also important, some applications requiring a higher degree of shrinkage than others. The amount of shrink can vary from 15 per cent to 80 per cent depending on the polymer composition and manufacturing techniques. Of particular interest is the relationship between degree of shrinkage and temperature, films with a steep shrink/temperature curve (*e.g.*, polypropylene, where a ±5°C variation in tunnel temperature could vary the degree of shrinkage by up to 20 percent) being more difficult to handle because of the closer temperature control necessary.

Shrink tension, the stress exerted by the film when it is restrained from shrinking at elevated temperatures, is the third important factor.

Generally, the lower the temperature at which orientation was carried out, the greater the shrink tension. Tension of 300 to 1000 kPa is desirable in order to provide a tight package after shrinking. With a shrink tension above 2000 kPa, care must be taken to prevent crushing and distorting of the package by limiting shrink temperature and time.

Balanced orientation is especially important for printed films, since uniform shrinkage is essential to avoid distortion of the print after shrinkage. Even a balanced biaxially oriented film may not shrink evenly in both directions if the product is of a very irregular shape. In these situations, it may be necessary to chose a print design which is not affected by such distortion.

The stretch films were first introduced in the early 1970s. In stretch wrapping the film is stretched around the article and the loose end 'wiped" against the underlying film, the film-to-film adhesion or cling being sufficient to hold it in place. Other less common ways of attaching the end are heat sealing, adhesives, mechanical fasteners and tying. Orientation of the polymer chains occurs on stretching to give a stiffer film, improving the tightness of the wrap and the stability of the load. The residual tension in the film gives a tight contour wrap. A simple analogy is that of an elastic or rubber band which can be stretched over an object and then assume a tight position.

The main films used in stretch wrapping are low density polyethylene, linear low density polyethylene, poly(vinyl chloride), ethylene-vinyl acetate copolymer and polypropylene, the choice depending on such factors as appearance (*i.e.,* requirements for clarity, sparkle, etc.) and the protection required (gas and moisture barrier, and/or physical protection in preventing pallet loads from disintegrating).

There are four important mechanical properties of stretch films. The stretchability or elongation is a measure of how much a given film can be stretched without distorting the load, the greater the stretchability, the lower the material costs since less film is used for a given application. The stress is a measure of the static force exerted by the film on the load, and is highest at the corners of a straight, square load. This force can crush or distort the load if not controlled during the wrapping process. The restretch force or elasticity is a measure of the force available to restrict further movement of the load during transit as a result of vibration, sway and high impact forces. The breaking strength is a measure of the ultimate force that the film can withstand without failure.

Low density polyethylene is classed as a low stretch film with a stretchability of about 30 per cent. Under ideal conditions linear low density polyethylene can stretch up to 400 per cent, although the practical limit is nearer 200 per cent. About 60 per cent of the initial stretch is retained by this film, the figure for low density polyethylene being about 70 per cent. Both are quite adequate for the stretch wrapping of large pallet loads. The trend is towards lighter gauge films, 20 μm being common, with some as thin as 13 μm.

An important property of stretch films is their cling with blown linear low density polyethylene having much less cling than poly(vinyl chloride) or ethylene vinyl acetate copolymer, although cast linear low density polyethylene approaches the others in this property. Additives such as glyceryl mono-oleate and polyisobutylene are sometimes added by film manufacturers to improve cling. Elevated humidity can sometimes enhance film cling because some cling additives function by attracting moisture from the atmosphere. As well, antioxidants, antistatic agents and antiblock agents (the latter to prevent the layers of film on a roll from becoming permanently bonded together) are also frequently added.

The advantages and disadvantages of stretch versus shrink film can be considered under various headings. From the energy point of view, stretch wrapping requires no shrink tunnel or hot air guns. As a consequence, stretch wrap systems can be used in cold stores. There are also savings in yield, since a thinner gauge film can be used for stretch wrapping. With a shrink film, typically 1 kg of film per apllet is used; this was reduced to 0.5 kg with the early stretch films. The new linear low density polyethylene films can be used at around 0.3 kg per pallet. Furthermore, stretch film will not stick to film-wrapped packs as will shrink film if used to overwrap pallet loads. Installation and commissioning costs are usually less for stretch film.

One early disadvantage of stretch wrapping was the lack of control of film tension, resulting in many instances in damage to the goods being wrapped due to excessive film tension. Now-a-days suitable equipment incorporating adjustable torque units is available to overcome this problem. One associated problem is that different films retain different stress, since all films start to relax immediately after they have been stretched. Most of the relaxation (about 99 per cent) occurs within 24 hours. The opposite of relaxation is stress retention which is defined as the capacity of a film to maintain the applied tension during stretch wrapping.

The percentage of *stress retention* should not be confused with the percentage of *stress recovery*, the latter being analogous to the behaviour of a rubber band. Stress recovery is important whenever the unit load decreases in size. Then stress relaxation is not due to film elongation with time, but to the reduction in the size of the load. In these situations, stretch wrapping has major advantages compared with most strapping methods since the latter provide very little stress recovery.

Blown low density polyethylene has about 85 per cent stress recovery while poly(vinyl chloride) has about 100 per cent, after 16 hours at 20 per cent initial elongation and 25°C. Thus, the former film would be more suitable for containing heavier loads with the latter being preferred when significant reduction in unit load size occurs. However, such a relationship may not hold at higher ambient temperatures since stress retention of PVC decreases much faster than LDPE as the temperature increases. Stress recovery must be taken into account when choosing a stretch film and setting the tension on the torque unit.

Recent developments include the introduction of pre-stretching, where stretching occurs between two rollers immediately prior to the film being applied to the load. This permits films to be stretched to a higher ratio and the weight of film per pallet is thus reduced. A further advantage with this system is that less 'necking' or narrowing of the film occurs as it is stretched, because the force is applied more evenly over a shorter distance.

An alternative approach to high stretch is the use of a pre-heat system where the film is heated before stretching, reducing the force required to stretch the film and thus, the risk of mechanical damage. A further development has been the introduction of power stretch systems where the driving force for stretch is not the rotating load but the driven rollers over which the film is drawn. The stretching force is thus isolated from the load which reduces the risk of film damage at high stretch.

9.5 The Types of Sealing

Heat Sealing

The heat sealability of a packaging film is one of the most important properties when considering its use, and the integrity of the resultant seal is of paramount importance to the ultimate package integrity.

A number of factors are involved in determining the quality of a heat seal. They can be conveniently summarized under three headings:

1. *Machine factors*: dwell or clamp time; temperature and pressure.
2. *Resin factors*: density; molecular weight and additives in the resin.
3. *Film factors*: gauge; style or form (*e.g.*, whether gussetted or not) and treatment for printing.

All of these factors tend to interact in a complex way. For example, the amount of heat available may be limited by the capacity of the heating elements, by the rate of heat transfer of the sealing bar and its coating, or the type of product being packaged. Increasing the dwell time (*i.e.*, the time during which heat is applied) will increase the heat available, but this may prove to be uneconomic since fewer units will be able to be handled per minute.

Heat sealable films are considered to be those films which can be bonded together by the normal application of heat, such as by conductance from a heavy heat-resistant metal bar containing a heating element. Nonheat-sealable films cannot be sealed this way, but they can often be made heat sealable by coating them with heat-sealable coatings. In this way, the two facing coated surfaces become bonded to each other by application of heat and pressure for the required dwell time.

Conductance Sealing

Conductance (also known as resistance) sealers are the most common type of heat sealers in commercial use and typically consist of two metal jaws (often patterned or embossed to give the seals extra strength), one of which is electrically heated, the temperature being controlled thermostatically. The second or backing jaw is often covered with a resilient material such as rubber to distribute pressure evenly and aid in smoothing out the film in the sealing area. Frequently the unheated jaw is water cooled, although in some situations it may be heated to the same temperature as the first jaw to enable sealing through sheets of film in exactly the same way each time.

Conductance sealers are used for coated regenerated cellulose films or any other materials with a heat seal coating such as foil or paper. However, they are not suitable for unsupported materials such as polyethylene film which would simply melt and stick to the jaws. Serrated jaws can be used to ensure that the two webs are stretched into intimate contact with high local pressure; they also improve appearance. For all sealing jaws a non-stick coating is desirable; Teflon-[poly(tetrafluoroethylene)] is commonly used, either as a cloth-reinforced film or, in the case of serrated jaws, as a powder bound in a heat-resistant vehicle.

Dwell time should be able to be controlled to fractions of a second and be easily adjustable. Likewise, the pressure between the jaws should also be easily adjustable. Both these factors will need to be changed when different materials are heat sealed.

A variation on this type of sealer is the band sealer, where the films travel between two endless bands of metal which are pressed together by heated bars. The heat passes through the bands and seals the films; the bands are then pressed together by chilled bars to withdraw heat from the seal. Band sealers are widely used for sealing pouches and have the advantage of being continuous.

Impulse Sealing

In this method the films (usually unsupported materials) are clamped between two cold metal bars and then fused by the effect of a short heat impulse; cooling occurs under pressure. Much lighter jaws than for resistance sealers are used. A high current is sent for a short period through a nichrome resistance wire or ribbon covered with Teflon tape. The current heats the wire to the desired temperature, the temperature being controlled with a transformer.

Dwell time of the heating impulse must also be controlled, as must the length of the cooling period. Thus, two timers are often found on impulse sealers, the second timer controlling the cooling

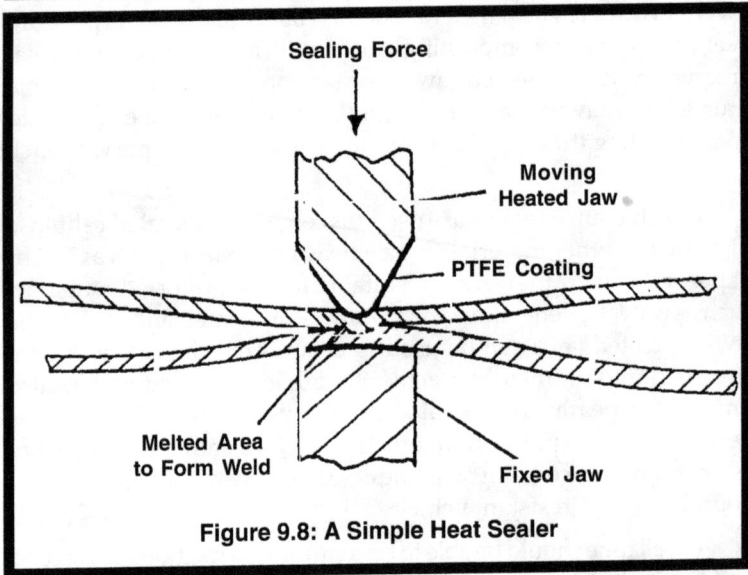

Figure 9.8: A Simple Heat Sealer

of the resistance wire to allow the film to harden under pressure and prevent deformation of the film in the sealed area. Sometimes one of the jaws is water cooled to prevent excess heating and promote rapid cooling. If very heavy films are to be sealed, both jaws may contain resistance wires. To prevent the film from sticking to the heated jaws, it is usual to cover them with a slip sheet such as glass cloth impregnated with Teflon. Generally, the seals produced by impulse sealing are of excellent quality.

Ultrasonic Sealing

In this method of sealing, a generator feeds a 20 kHz signal into a transducer which transforms electrical energy into mechanical vibrations of the same frequency. The mechanical energy is converted into heat at the interface and produces an almost instantaneous weld. Little overall heating is produced, thus enabling oriented films to be sealed without any change in dimensions. This method is also useful for sealing multiwall paper bags which have an inner polyethylene liner, since the heat is generated only at the welding interface and there is no heat damage to the paper.

A conical welding head acts as a focussing tool for the vibrations which pass through the film to an anvil. The pressure, along with the dwell time (expressed in terms of the rate at which the film is fed between the welding head and anvil) and temperature, is adjusted to get optimum seal quality.

Dielectric Sealing

In this method, dielectric heating is used in which a high-frequency current (typically 50–80 MHz) is passed through two or more layers of film. The electrodes (usually made of brass) are the top and bottom jaws, the latter being the ground of the circuit. When the layers of film are in place between the closed jaws, a high-frequency current is passed between the jaws, heating and liquefying the films. The pressure which the jaws exert on the films helps to bring about thorough fusion and bonding.

This method of sealing is used principally with poly(vinyl chloride) and nylon 6/6 films as they are difficult to heat seal by direct means because they tend to degrade at temperatures close to their softening point. It is only applicable to materials which are polar and capable of forming a dipole moment.

Hot-wire Sealing

This method of sealing (which uses a thin wire or strip of metal with a, radiussed edge, heated with a low voltage current) has found application for the manufacture of polyethylene bags and pouches in tubular form, the cutting and sealing operations being carried out in the one step. It is only applicable to thermoplastic films which can tolerate high temperatures for a short time and which also have a low viscosity in the fused stage. When unsupported films are trim sealed by this method, they tend to form a strong bead in their seal areas due to surface tension and orientation. This method is also used to a limited degree with laminated constructions.

There is a tendency with this method of sealing to get what is known as "angel hair"–fine strands of polymer protruding from the seal area. This can be controlled with the use of proper temperatures and times. Generally, films thicker than 0.05 mm are difficult to seal through, especially if they have gussetted structures.

Chapter 10
Use of Nanotechnology in Foods and their Packaging

Introduction

According to a definition in a recent report (Nanotechnology in Agriculture and Food) food is "nanofood" when nanoparticles, nanotechnology techniques or tools are used during cultivation, production, processing, or packaging of the food. It does not mean atomically modified food or food produced by nanomachines.

Nanotechnology is an exciting and rapidly emerging technology allowing us to manipulate and create materials and structures at the molecular level, often atom by atom into functional structures having nanometer dimensions. Nanotechnology enables the designers to alter the structure of the packaging materials on the molecular scale, to give the materials desired properties. Simple traditional "packing" is to be replaced with multi-functional intelligent packaging methods to improve the food quality, thanks to the application of nanotechnology in this field. This will make products cheaper, production more efficient and more sustainable through using less water and chemicals. Producing less waste and using less energy is a central concern of food manufacturers, and the drive towards production efficiency is likely to continue to boost nanotechnology funding (Asadi and Mousavi, 2007).

Nanoscale biotech and nano-bio-info will have big impacts on the food and food-processing industries. The future belongs to new products, new processes with the goal to customize and personalize the products. More than 180 applications are in different developing stages and a few of them are on the market already. The nanofood market is expected to surge from 2.6 bn. US dollars today to 20.4 bn. US dollars in 2010. More than 200 Companies around the world are today active in research and development. USA is the leader followed by Japan and China. By 2010 Asian with more than 50 per cent of the world population will be the biggest market for Nanofood with the leading of China (Moraru *et al.*, 2003).

Application of Nanotechnology in Food Science

The potential benefits of Nanofoods–foods produced using nanotechnology are astonishing. Nanotechnology is having an impact on several aspects of food science, from how food is grown to how it is packaged. Advocates of the technology promise improved food processing, packaging and safety; enhanced flavor and nutrition; 'functional foods' where everyday foods carry medicines and supplements, and increased production and cost-effectiveness. Companies are developing nanomaterials that will make a difference not only in the taste of food, but also in food safety, and the health benefits that food delivers.

In a world where thousands of people starve each day, increased production alone is enough to warrant worldwide support. For the past few years, the food industry has been investing millions of dollars in nanotechnology research and development. Some of the world's largest food manufacturers, including Nestle, Altria, H.J. Heinz and Unilever, are blazing the trail, while hundreds of smaller companies follow their lead. Yet, despite the potential benefits, compared with other nanotechnology arenas, nanofoods don't get a lot of publicity. The ongoing debate over nanofood safety and regulations has slowed the introduction of nanofood products, but

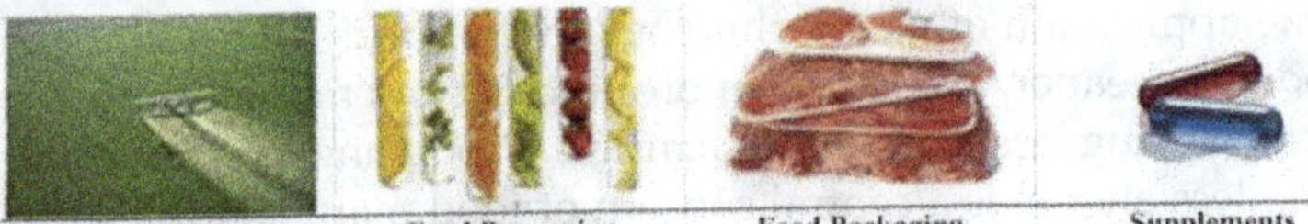

Agriculture	Food Processing	Food Packaging	Supplements
• Single molecule detection to determine enzyme/ substrate interactions • Nanocapsules for delivery of pesticides, fertilizers and other agrichemicals more efficiently • Delivery of growth hormones in a controlled fashion • Nanosensors for monitoring soil conditions and crop growth • Nanochips for identity preservation and tracking • Nanosensors for detection of animal and plant pathogens • Nanocapsules to deliver vaccines • Nanoparticles to deliver DNA to plants (targeted genetic engineering)	• Nanocapsules to improve bioavailability of neutraceuticals in standard ingredients such as cooking oils • Nanoencapsulated flavor enhancers • Nanotubes and nanoparticles as gelation and viscosifying agents • Nanocapsule infusion of plant based steroids to replace a meat's cholesterol • Nanoparticles to selectively bind and remove chemicals or pathogens from food • Nanoemulsions and –particles for better availability and dispersion of nutrients	• Antibodies attached to fluorescent nanoparticles to detect chemicals or foodborne pathogens • Biodegradable nanosensors for temperature, moisture and time monitoring • Nanoclays and nanofilms as barrier materials to prevent spoilage and prevent oxygen absorption • Electrochemical nanosensors to detect ethylene • Antimicrobial and antifungal surface coatings with nanoparticles (silver, magnesium, zinc) • Lighter, stronger and more heat-resistant films with silicate nanoparticles • Modified permeation behavior of foils	• Nanosize powders to increase absorption of nutrients • Cellulose nanocrystal composites as drug carrier • Nanoencapsulation of neutraceuticals for better absorption, better stability or targeted delivery • Nanocochleates (coiled nanoparticles) to deliver nutrients more efficiently to cells without affecting color or taste of food • Vitamin sprays dispersing active molecules into nanodroplets for better absorption

Figure 10.1: Examples for Nanofood Applications
(*Source*: Nanowerk)

research and development continue to thrive - though, interestingly, most of the larger companies are keeping their activities quiet. Although the risks associated with nanotechnology in other areas, such as cosmetics and medicine, are equally blurry, it seems the difference is that the public is far less apt to jump on the nanotechnology bandwagon when it comes to their food supply.

In the forefront of nanofood development is Kraft Foods, which took the industry's lead when it established the Nanotek Consortium, a collaboration of 15 universities and national research labs, in 2000. Kraft's focus is on "interactive" foods and beverages. These products will be customized to fit the tastes and needs of consumers at an individual level. Possible products include drinks that change colors and flavors to foods that can recognize and adjust to a consumer's allergies or nutritional needs. Other large companies, such as Nestlé and Unilever, are exploring improved emulsifiers that will make food texture more uniform. These huge Western companies are responsible for the bulk of the food industry's research and development, however the nanofood industry is truly a global phenomenon.

Nanoparticles are being developed that will deliver vitamins or other nutrients in food and beverages without affecting the taste or appearance. These nanoparticles actually encapsulate the nutrients and carry them through the stomach into the bloodstream. In Australia for instance, nanocapsules are used to add Omega-3 fatty acids to one of the country's most popular brands of white bread. According to the manufacturer, nanocapsules of tuna fish oil added to Tip Top Bread provide valuable nutrients, while the encapsulation prevents the bread from tasting fishy. NutraLease, a start-up company of the Hebrew University of Jerusalem has developed novel carriers for nutraceuticals in food systems. The nano-sized self-assembled structured liquids (NSSL) technology allows for encapsulation of nutraceuticals, cosmeceuticals, and essential oils and drugs in food, pharmaceuticals, and cosmetics. Another advantage to the NSSL technology is that it allows the addition of insoluble compounds into food and cosmetics. One of the first products developed with this technology, a healthier version of canola oil is already available to consumers in Israel.

In other parts of the world, nanotechnology efforts are focused on the agricultural side of food production. A joint effort among universities in India and Mexico is directed at developing non-toxic nanoscale herbicides. Researchers at Tamil Nadu Agricultural University in India and Monterrey Tech in Mexico are looking for ways to attack a weed's seed coating and prevent it from germinating.

The range of current nanofood research and development is as impressive as the industry's projected growth. Last August, UK-based Cientifica estimated that nanotechnologies in the food industry were currently valued at $410 million and would grow to $5.8 billion by 2012.

Application of Nanotechnology in Food Packaging

Today, food-packaging and monitoring are a major focus of food industry-related nanotech R&D. Novel food packaging technology is by far the most promising benefit of nanotechnology in the food industry in the near future. Nanotechnology involves the study and use of materials at an extremely small scale at sizes of millionths of a millimetre and exploit the fact that some materials have different properties at this ultra small scale from those at a larger scale. One nanometer is the same as one millionth of a millimetre. The new nanocomposites developed by Ding and his team are assemblies of functionalised nanoparticles, hundreds of micrometres in size, capable of disintegrating in liquid into nanoparticles that attach to and kill microorganisms. Many scientists believe that such discoveries are pointing the way to what packaging will be like in the future. The Foundation for Scientific and Industrial Research at the Norwegian Institute of Technology (SINTEF) for example is already using nanotechnology to create small particles in the film and improve the transportation of some gases through the plastic film to pump out dirty air such as carbon dioxide. It is hoped that the concept could be used to block out harmful gases that shorten the shelf life of food. SINTEF scientists are looking at whether the film could also provide barrier protection and prevent gases such as oxygen and ethylene from deteriorating food.

Likewise, the team at the University of Leeds is hopeful that nanoparticle zinc oxide and magnesium oxide could provide safe and affordable food packaging in the not-too-distant future. According to packaging consultancy Pira, the team plans to incorporate effective antimicrobial nanoparticles into packaging materials as research progresses.

Much of this technology will require years of further development before it can be used commercially, but advancements in computer technology could provide a short cut. Computers are constantly becoming more powerful and capable of conducting more detailed explorations, and at the same time, scientists across the globe are increasingly becoming interested in the potential of nanotechnology.

It is the intersection of these two trends that is allowing scientists such as Ding, Povey and Zhang to investigate realms that are too small for today's technology to explore experimentally. The US nanomaterial market alone, which totalled only $125 million in 2000, is expected to reach $1.4 billion in 2008 and exceed $30 billion by 2020, according to Nanomaterials, a study from the Freedonia Group.

Packaging that incorporates nanomaterials can be "smart," which means that it can respond to environmental conditions or repair itself or alert a consumer to contamination and/or the presence of pathogens. According to industry analysts, the current US market for "active, controlled and smart" packaging for foods and beverages is an estimated $38 billion and will surpass $54 billion by 2008 (Donald, 2004).

Companies are already producing packaging materials based on nanotechnology that are extending the life of food and drinks and improving food safety. A UK research institute believes it has identified safe and effective antimicrobial nanoparticles for food packaging, a discovery that could revolutionise how food is packaged in the future (Ding and Povey, 2005). Professors Yulan Ding and Malcolm Povey and PhD candidate Lingling Zhang are confident that nanoparticles of zinc oxide and magnesium oxide have been shown to be effective in killing microorganisms. This could provide a cheap, safe alternative to nano-sized silver, which has good antimicrobial properties, but is expensive and as a heavy metal, is not suitable for human contact.

While the nanofood industry struggles with public concerns over safety, the food packaging industry is moving full-speed ahead with nanotechnology products. Today, food packaging and monitoring are a major focus of food industry-related nanotech R&D. Packaging that incorporates nanomaterials can be "smart," which means that it can respond to environmental conditions or repair itself or alert a consumer to contamination and/or the presence of pathogens.

Numerous companies and universities are developing packaging that would be able to alert if the packaged food becomes contaminated; respond to a change in environmental conditions; and self-repair holes and tears. One of the most promising innovations in smart packaging is the use of nanotechnology to develop antimicrobial packaging. Scientists at big name companies including Kraft, Bayer and Kodak, as well as numerous universities and smaller companies, are developing a range of smart packaging materials that will absorb oxygen, detect food pathogens, and alert consumers to spoiled food. These smart packages, which will be able to detect public health pathogens such as salmonella and *E. coli*, are expected to be available within the next few years. The following examples illustrate nanoscale applications for food and beverage packaging.

Using Clay Nanoparticles to Improve Plastic Packaging for Food Products

Use of nanomaterials in food packaging is already a reality. Clay nanocomposites are being used to provide an impermeable barrier to gases such as oxygen or carbon dioxide in lightweight bottles, cartons and packaging films. Clay nanocomposites are being used in plastic bottles to extend the shelf life of beer and make plastic bottles nearly shatter proof.

Chemical giant Bayer produces a transparent plastic film (called Durethan) containing nanoparticles of clay. The nanoparticles are dispersed throughout the plastic and are able to block oxygen, carbon dioxide and moisture from reaching fresh meats or other foods. The nanoclay also makes the plastic lighter, stronger and more heat-resistant.

One example is bottles made with nanocomposites that minimize the leakage of carbon dioxide out of the bottle; this increases the shelf life of carbonated beverages without having to use heavier glass bottles or more expensive cans.

With different nanostructure, the plastics can obtain various gas/water vapor permeability to fit the requirements of preserving fruit, vegetable, beverage and other foods. By adding nanoparticles, people can also produce bottles and packages with more light resistance, stronger mechanical and thermal performance, and less gas absorption. These properties can significantly increase the shelf life, efficiently preserve flavor and colour, and facilitate transportation and usage. Further, nanostructured film can effectively prevent the food from the invasion of bacteria and microorganism and ensure the food safety.

Another example is food storage bins with silver nanoparticles embedded in the plastic. The silver nanoparticles kill bacteria from any food previously stored in the bins, minimizing harmful bacteria.

Nanotechnology in Plastics Packaging

The study published by Pira (2004) takes a detailed look at what these break-through technologies mean for plastics packaging. It looks at how the usability, durability and value-added of plastics packaging materials can be enhanced through nanotechnologies such as silicate nanoparticles, carbon nanofibres, carbon nanotubes, electrospun nanotubes, nanocapsules and self-assembled monolayers. It looks at all current and future nano-related research projects or products in these technologies or applications that could have an impact on the plastics packaging sector.

Nanoscale titanium dioxide particles could be the next breakthrough in plastic packaging, having the ability to prevent ultraviolet light from reducing shelf life.

Over time the sun's ultraviolet light can cause cracking, fading and other types of solar degradation to plastics. Ultraviolet light (UV) can also reduce the shelf life and effectiveness of products in transparent or semi-transparent packaging.

Finding better ways to block UV rays from passing through plastic packaging, while still allowing consumers to see the product inside, is a major goal within the industry (Elamin and Bird, 2007).

DuPont announced the release of its Light Stabilizer 210, a plastic additive designed using extremely small particles of titanium dioxide, which have been nanoengineered to absorb UV. He lists current potential applications for Light Stabilizer 210 as consumer packaged products; the company has also applied in the US for regulatory approval for use in the food sector. In testing, Light Stabilizer 210 blocked twice as much UV light as several classes of competitive products, the company claimed in a statement.

Light Stabilizer 210 works by absorbing UV rays and changing them into small amounts of heat which dissipate quickly without damaging the structure of plastic, DuPont stated. The key performance advantage of the new light stabiliser is that its extremely small particle size provides much more surface area for UV absorption, as a sizeable percentage of titanium dioxide particles in the product are nanoscale.

Creating a Molecular Barrier by Embedding Nanocrystals in Plastic for Improving Packaging

Until recently, industry's quest to package beer in plastic bottles (for cheaper transport) was unsuccessful because of spoilage and flavour problems. Today, Nanocor, a subsidiary of Amcol International Corp., is producing nanocomposites for use in plastic beer bottles that give the brew a six-month shelf-life. By embedding nanocrystals in plastic, researchers have created a molecular barrier that helps prevent the escape of oxygen. Nanocor and Southern Clay Products are now working on a plastic beer bottle that may increase shelf-life to 18 months. Zinc oxide nanoparticles can be incorporated into plastic packaging to block UV rays and provide anti bacterial protection, while improving the strength and stability of the plastic film.

Using Nanotechnology Methods to Develop Antimicrobial Packaging and 'Active Packaging'

Kodak, best known for producing camera film, is using nanotech to develop antimicrobial packaging for food products that will be commercially available in coming time. Kodak is also developing other 'active packaging,' which absorbs oxygen, thereby keeping food fresh.

Using a Nanotech Bioswitch in 'Release on Command' Food Packaging

Researchers in the Netherlands are going on further to develop intelligent packaging that will release a preservative if the food within begins to spoil. This "release on command" preservative packaging is operated by means of a bioswitch developed through nanotechnology.

Embedded Sensors in Food Packaging and 'Electronic Tongue' Technology

Scientists at Kraft, as well as at Rutgers University and the University of Connecticut, are working on nano-particle films and other packaging with embedded sensors that will detect food pathogens, called "electronic tongue" technology, the sensors can detect substances in parts per trillion and would trigger a colour change in the packaging to alert the consumer if a food has become contaminated or if it has begun to spoil (Anonymous, 2004). With embedded nanosensors in the packaging consumers will be able to "read" the food inside. Sensors can alarm us before the food goes rotten or can inform the exact nutrition status contained in the contents. Nanotechnology is going to change the fabrication of the whole packaging industry. Self-assembly will reduce the fabrication costs and infrastructure. More flexible packaging methods will provide the consumers with fresher and customized products (Asadi and Mousavi, 2007).

Using Food Packaging Sensors in Defence and Security Applications

Developing small sensors to detect food-borne pathogens will not just extend the reach of industrial agriculture and large-scale food processing. In the view of the US military, it's a national security priority. With present technologies, testing for microbial food-contamination takes two to seven days and the sensors that have been developed to date are too big to be transported easily. Several groups of researchers in the US are developing biosensors that can detect pathogens quickly and easily, reasoning that "super sensors" would play a crucial role in the event of a terrorist attack on the food supply (Moraru et al., 2003). With USDA and National Science Foundation funding, researchers at Purdue University are working to produce a hand-held sensor capable of detecting a specific bacteria instantaneously from any sample. They've created a start-up company called BioVitesse.

Regulating Nanotech in Food Industry

While there are lots of opportunities for using nanotechnology to improve food production, packaging, and quality, there is also some concern about how this will play out. For example the organizers of the Joint Symposium on Food Safety and Nutrition, organized by the Central Science Laboratory in the UK and the Joint Institute for Food Safety and Applied Nutrition at the University of Maryland, had chosen to focus their 2007 symposium on Nanotechnology in Foods and Cosmetics. They feel that nanotech materials both have "the potential for use in a vast variety of products and may pose new and unique safety issues".

According to the research, nanotech in food packaging will bring benefits in the form of:

1. Reduced usage of energy and raw materials (polymers)
2. Improvement of the mechanical properties of biodegradable polymers
3. Beneficial for reusable packaging due to improved mechanical strength and
4. Help to reduce the amount of packaging waste and save resources at the same time.

There are however, risks associated with this. The first type of the risk deals with inconclusive research on impacts and lack of effects data:

1. Knowledge about the effects of engineered nano-materials in the environment is still insufficient

2. Risk of environmental pollution by nanoparticles (persistence and unknown catalytic effects) whereas the rests of the risks are common risks also found in normal chemicals which relates to pollution and waste:

   (*a*) Release of nano particles into the environment during the production and disposal phases of a product,

   (*b*) Uncontrolled incineration of nanocomposites could cause emissions of nanoparticles into environment and

   (*c*) Nanocomposite materials could disturb plastic recycling processes. Recyclers will have to deal with nanoparticular fillers, which eventually will be found in recycled materials.

In its February, 2007 meeting, the European Food Safety Authority Regulatory agency announced that it was forming a scientific panel to conduct a risk assessment of nanoparticles in food and food packaging. This panel should be able to draw input and expertise from across Europe. For example, Denmark's National Food Institute is working on a project to gather toxicology information on nanoparticles and the UK Food Safety Authority has put together a report that provides "an outline of potential areas for future regulation relating to the use of nanotechnology and nanomaterials in foods".

In August 2006, the U.S. Food and Drug Administration (FDA) formed a Nanotechnology Task Force with goals that included:

1. Evaluate the effectiveness of the agency's regulatory approaches and authorities to meet any unique challenge that may be presented by the use of nanotechnology materials in FDA-regulated products.

2. Explore opportunities to foster innovation using nanotechnology materials to develop safe and effective drugs, biologics, and devices, and to develop safe foods, feeds and cosmetics.

While the regulatory agencies may be making these efforts a little late, because some products are already available and development has been started on many more, one can hope that discussions would help consumers to benefit from improved and safe food products with a minimum of controversy (Earl Boysen).

Activist Concerns

Countering these developments are civil organisations such as the ETC group, which has expressed general concerns about the use of nanotechnology in food and agriculture and scrutinises socio-economic effects. In a report published last year the ETC group called for a moratorium, prohibiting the application of nanotechnology until the absence of health risks caused by nanoparticles will be scientifically evident.

This is a major warning sign that nanotechnology may soon hit a consumer roadblock once products come on the market. The report is part of a series of outlines being published this year as the European Commission develops a ten-year strategic plan for using the technology. More reports are expected to be published by the end of this year, the deadline.

The OCA (Organic Consumers Association), a grassroots non-profit public interest organization based in the U.S., is one of many vocal organizations calling for government regulation on nanofoods, at least until more safety testing is completed. These organizations argue that a lack of evidence of harm is not the same as reasonable certainty of safety, which is what food companies must demonstrate to the U.S. Food and Drug Administration (FDA) before introducing a new food additive.

"The OCA is focusing its efforts on educating the public about the potential risks of nanofoods and putting pressure on government agencies to increase oversight," says Minowa, adding that ever-tightening federal budgets, at least in the U.S., will make the latter a huge challenge. "There's a lack of consumer understanding, a lack of government oversight and a lack of labeling," says Minowa. "Combine these with a lack of testing and you have an equation for serious problems."

Although there is far less opposition to nanopackaging than there is to nanofoods, there are some who argue that the use of these devices will allow the food industry to further shirk their corporate responsibilities.

"While devices capable of detecting food-borne pathogens could be useful in monitoring the food supply, sensors and 'smart packaging' will not address the root problems inherent in industrial food production that result in contaminated foods: faster meat (dis)assembly lines, increased mechanization, a shrinking labor force of low-wage workers, fewer inspectors, the lack of corporate and government accountability and the great distances between food producers, processors and consumers," says the ETC Group, a conservation and sustainable advancement organization. "Just as it has become the consumer's responsibility to make sure meat has been cooked long enough to ensure that pathogens have been killed, consumers will soon be expected to act as their own meat inspectors so that industry can continue to trim safety overhead costs and increase profits."

Interestingly enough, the Environmental Protection Agency (EPA) in the U.S. declared on November 22, 2006 that it intends to regulate a large class of consumer items made with silver nanoparticles. The decision, which will affect not only washing machines but other consumer products such as odour-destroying shoe liners, food-storage containers, air fresheners, and a wide range of other products that contain nanosilver, marks a significant reversal in federal policy. Nanosilver containing consumer products that are applied to food packaging are not regulated by the EPA but by the FDA. The FDA is still considering whether it needs new rules for nanomaterials (Garber, 2007).

Chapter 11

Packaging of Cereals

11.1 Introduction

Cereals are the fruits of cultivated grasses, members of the monocotyledonous family *Gramineae.* The principal cereal crops are wheat, barley, oats, rye, rice, maize, sorghum and the millets. Cereals have been important crops for thousands of years and the successful production, storage and use of cereals has contributed in no small measure to the development of modern civilization. Today cereals and cereal-based products are an important part of the diet in most countries, and each year new products based on cereals are developed and marketed to increasingly sophisticated consumers.

The cereals of commerce and industry are harvested, transported and stored in the form of grain. The anatomical structure of all cereal grains is basically similar, differing from one cereal to another in detail only.

The mature grain of the common cereals consists of carbohydrates, nitrogenous compounds (mainly proteins), lipids, mineral matter and water, together with small quantities of vitamins, enzymes and other substances, some of which are important nutrients in the human diet. Carbohydrates are quantitatively the most important constituents, forming 77–87 per cent of the total dry matter. The lipids in milled cereal products are liable to undergo two types of deterioration: hydrolysis from endogenous lipases, and oxidation from endogenous lipoxygenases or molecular oxygen. The products of lipid hydrolysis are glycerol and free fatty acids which give rise to unpleasant odors. The products of lipid oxidation cause the odor and flavor of rancidity. Damage to the grain and the fragmentation that occurs in milling promote deterioration by bringing the lipid and the enzyme together.

The processing of cereals for the manufacture of food may bring about alteration in the chemical composition in a number of ways:

1. Parts of the grain may be separated during processing and removed from the product, *i.e.*, the product may constitute only a fraction of the grain.

2. The various nutrients may be distributed non-uniformly throughout the various parts of the grain, so that certain nutrients are preferentially lost from or concentrated into the products when separation is made.

3. The processing treatments may bring about changes in the nutrients themselves. For example, chemical changes such as inactivation of enzymes or hydrolysis of polysaccharides, and changes in distribution such as translocation of vitamins that occurs during the parboiling of rice.

The hazards to grain in storage are moisture, temperature, fungi, bacteria, insects and other pests. If the grain moisture content can be controlled, the hazards due to temperature-rise, fungi and insects can be largely avoided.

It has been known for many years that it is not the moisture content *per se* that is important for long term storage but rather the water activity a_w, a "safe" moisture content for long term storage of grain and oil-seeds usually being accepted as one in equilibrium with 70 per cent relative humidity, *i.e.*, 0.70 a_w. Above 0.75 a_w molds will develop rapidly during storage and heating will occur, with subsequent deterioration in, and loss of, product. Moisture sorption isotherms for wheat, barley and maize have been published and all have a typical sigmoid shape. The moisture content corresponding to $0.70\,a_w$ varies quite considerably among the different grains, thus emphasizing the utility of expressing stability in terms of a_w rather than moisture content.

Storage facilities for grains take many forms, ranging from piles of unprotected grain on the ground, underground pits or containers, and piles of bagged grain, to storage bins of many sizes, shapes and types of construction. It is essential that the grain has been dried to a moisture content corresponding to $0.70\,a_w$ or less prior to packaging and storage. Consumer packages for grain commonly consist of heat sealed pouches made from LDPE film; these provide a satisfactory moisture barrier and result in the required shelf lives for the grains.

11.2 The Storage of Wheat and Rice

The wheat grain is a living, respiring organism usually carrying endemic fungi. Respiration is slow at 14 per cent moisture content and 20°C, but rises as moisture content and temperature rise. The process of respiration generates heat (which is difficult to remove since wheat is a poor conductor) as well as CO_2 and water vapor, resulting in a loss in weight. Unless the grain is turned over to allow evaporation of the water, it will sweat and become caked in the bin.

Wheat at moisture contents between 16 per cent and 30 per cent can support fungal growth and there is thus, the risk of mycotoxin production. Above 30 per cent moisture content, wheat is susceptible to bacterial attack, leading to spoilage, excessive heat production and possibly charring. Insect life also becomes more active as the temperature rises and, because of their respiration, live insects in grain also raise the grain temperature. Deterioration during storage is aggravated by mechanical damage during harvesting since microorganisms attack damaged grains more readily than intact grains.

The shelf life for wheat as a function of moisture content and temperature is presented in Table 11.1. This table shows the importance of drying grain in that a drop of 3 per cent in moisture content increases the shelf life by a factor of four. The major problems at these higher moisture contents include accelerated wheat enzyme activities and microbial spoilage.

Table 11.1: Safe Storage Life (Days) for Grains as a Function of Moisture Content and Temperature

Grain Temperature (°C)	Grain Moisture, %		
	14	15.5	17
10.0	256	128	64
15.5	128	64	32
21.1	64	32	16
26.7	32	16	8
32.2	16	8	4
37.8	8	4	2

Rice is the staple cereal in Japan and is also consumed in large quantities in many other countries. Brown (unmilled) rice is more nutritious than milled rice, but storage stability problems and a traditional consumer preference for whole (milled) rice have limited the quantities of brown rice packaged and sold for direct consumption. A major deterrent to greater user of brown rice is the accumulation of free fatty acids in rice stored under warm and humid conditions. Fatty acids can be released by lipase activity present in the rice aleurone (bran) layer of damaged grains and by high lipase-containing bacteria and fungi adhering to rice.

In a study to evaluate the effects of CO_2 gas flushing on the shelf life of brown rice, greater stability of the rice was obtained at refrigerated (4°C) storage temperatures when it was packaged in a laminate film (nylon-EVA Copolymer) bag rather than a regular (unspecified) plastic bag; gas flushing with CO_2 of the samples in the laminate bag improved stability even further. However, no differences between the three types of bags were found when stored at room (24°C) temperature. The gas flushed bags formed a hard, rigid pack similar to a vacuum packaged product and maintained appearance for at least 3 years. It was suggested that if bulk rice in warehouses could be stored as brown rice (minus hulls) instead of as rough rice (with hulls) in laminate, gas flushed bags, savings of at least 20 per cent by weight and 30–35 present by volume should result.

11.3 Breakfast Cereal Foods

Breakfast cereal foods can be classified according to:

1. The amount of domestic cooking required,
2. The form of the product, and
3. The cereal used as raw material.

All cereals contain a large proportion of starch which in its natural form is insoluble, tasteless and unsuited for human consumption. It must be cooked to make it digestible and acceptable. In the case of hot cereals the cooking is carried out in the home, while ready-to-eat cereals are cooked during manufacture.

If the cereal is cooked with excess water and moderate heat as in boiling, the starch gelatinizes and becomes susceptible to starch-hydrolyzing enzymes in the human digestive system. If the cereal is cooked with a minimum of water (or without water) but at higher temperatures as in toasting, nonenzymic browning between protein and reducing sugars may occur and there may be some depolymerization of the starch.

Ready-to-eat cereals probably owe their origin to the Seventh Day Adventist Church whose members, preferring an entirely vegetable diet, experimented with the processing of cereals in the mid-nineteenth century. A granulated product called "Granula" and made by J.C. Jackson in 1863 may have been the first commercially available ready-to-eat breakfast cereal. A similar product called "Granola" was made by J.H. Kellogg by grinding biscuits made from wheatmeal, oatmeal and maize meal.

Ready-to-eat cereals comprise flaked, puffed, shredded and granulated products, generally made from wheat, maize or rice, although oats and barley are also used. The basic cereal may be enriched with sugar, honey or malt extract. All types are prepared by processes which tend to cause hydrolysis (dextrinization) rather than gelatinization of the starch.

Flaked products are made from wheat, corn, oats or rice. After cooking (often at elevated pressure) and the addition of flavorings such as malt, sugar and salt the cereal is dried to 15–20 per cent moisture content and conditioned for 1–3 days. It is then flaked, toasted, cooled and packaged.

Puffed products are prepared from conditioned whole grain wheat, rice, oats or pearl barley or a dough made from corn meal or oat flour with the addition of sugar, salt and sometimes oil. It is cooked for 20 min under pressure, dried to 14–16 per cent moisture content and pelleted by extrusion through a die. A batch of the conditioned grain or pelleted dough is fed into a heated pressure chamber which is injected with steam. The starch becomes gelatinized and expansion of water vapor on release of the pressure causes a several-fold increase in volume. The puffed product is dried to 3 per cent moisture content by toasting and then cooled and packaged.

Shredded products are made from whole wheat grains which are cooked to gelatinize the starch. After cooling and conditioning, the grain is fed through shredders. The shreds are baked for 20 min at 260°C, dried to 1 per cent moisture content, cooled and packaged.

Granulated products are made from a yeast dough consisting of wheat flour and salt. The dough is baked as large loaves which are then broken up, dried and ground to a standard degree of fineness.

Flaked or puffed cereals are sometimes coated with sugar or candy to provide a hard, transparent coating that does not become sticky even under humid conditions. The sugar content of cornflakes increases from 7 per cent to 43 per cent as a result of the coating process, and that of puffed wheat from 2 per cent to 51 per cent.

Deterioration

There are five modes of deterioration to be considered when selecting suitable packaging materials for breakfast cereals. They are:

1. Moisture gain resulting in loss of crispness.
2. Lipid oxidation resulting in rancidity and off-flavors.
3. Loss of vitamins.
4. Breakage, resulting in an aesthetically undesirable product.
5. Loss of aroma from flavored product.

The shelf life of breakfast cereals depends to a large extent on the content and quality of the oil contained in them. Thus products made from cereals having a low oil content such as wheat, barley, rice and maize grits (oil content: 1.5–2.0 per cent) have a longer shelf life than products made from oats (oil content 4–11 per cent, average 7 per cent). Although whole corn has a high oil content (4.4 per cent), most of the oil is contained in the germ which is removed in making grits.

Packaging

A detailed description of the materials used for the packaging of ready-to-eat breakfast cereals has been provided by Monahan, and a brief summary of the essential aspects is given below.

Loss of Crispness

Data on the permissible increase in moisture content or a_w before loss of crispness occurs are scant and without this data, it is difficult to specify precisely the type of moisture vapor barrier required to achieve a given shelf life.

Packaging of breakfast cereals has traditionally been in fiberboard boxes with a supercalendered waxed glassine liner. More recently the glassine liner has been replaced by various plastic materials, in particular thin gauge HDPE which is usually folded rather than heat sealed. HDPE coextruded with a thin layer of EVA copolymer is a recent introduction, the EVA copolymer 'permitting a lower heat seal temperature and offering the consumer an appealing and peelable seal.

In a few cases where the cereal product is not hygroscopic and/or retains a satisfactory texture when in equilibrium with the ambient atmosphere, a liner may not be needed for moisture protection and may even serve to entrap rancid aromas. Where this is the case, either no liner or one which is vapor permeable may be used. Some shredded wheat products are in this category and are discussed below.

As well as providing a barrier to moisture vapor, the liner must also confine cereal aromas within and at the same time prevent foreign odors entering the packaged product. It should also be reclosable to protect the cereal remaining in the package.

Lipid Oxidation

The primary mode of chemical deterioration in dry cereals is lipid oxidation and two reasons have been advanced for this. First, the a_w of dry cereals is at or below the monolayer which essentially stops all other types of deteriorative reactions. Second, unsaturated fats are required in lipid oxidation and the grains used in breakfast cereals have a high ratio of unsaturated to saturated fats.

To minimize oxidative rancidity it is important that the package exclude light. Excluding oxygen may be of limited assistance in extending shelf life although oxygen is almost never rate limiting. For this reason, most companies do not bother to use packaging which is a good oxygen barrier. In a study of the storage stability of a flaked oat cereal product packaged in materials of different oxygen barrier properties with and without the addition of an iron-based oxygen absorber, the absorber retarded or delayed lipid oxidation provided that it was used with a packaging material such as PVC/PVdC copolymer-coated PP-LDPE that was a good oxygen barrier.

The use of an anti-oxidant in the package liner has been shown to be successful in extending shelf life but is not generally permitted in most countries.

In shredded cereals, rancid odors tend to accumulate if shredded wheat is stored in an airtight container, due to the higher unsaturated fat content in this cereal. Therefore, the product is usually sold in breather boxes without inner or outer linings. When so packaged, the product is just as stable as any other prepared cereal, except that moisture absorption can take place much more readily. The net effect is a reduction in shelf life of these products.

Loss of Vitamins

The vitamin and mineral fortification of cereals is widely practiced in many countries and there are usually associated nutritional labeling requirements. The major factor influencing vitamin loss in packaged cereals is the temperature of storage. In a study on the effects of processing and storage on, micronutrients in breakfast cereals, it was concluded that micronutrient loss would not be a major factor in determining the shelf life of dry cereals. There was no substantial losses of added vitamins during normal shelf lives, with the possible exception of vitamin A and, to a slight extent, vitamin C. Vitamin A survived six months (the average distribution time) at room temperature with no measurable loss.

Mechanical Damage

The rigidity of the carton stock and the compression resistance of the finished carton together must provide the necessary resistance to product breakage throughout production line operations, warehouse storage and distribution from the manufacturer to the retailer and consumer. Rigidity also prevents the bulging of the carton. Protecting breakfast cereals from breakage does not appear to be a problem using currently available carton stock and carton designs.

Loss of Flavor

This can be a problem with certain cereal products to which fruit flavors have been added prior to packaging. In these situations, loss of flavor results in the product being considered to be at the end of its shelf life by the consumer. A study evaluating two typical cereal liner materials (HDPE and glassine) found that the permeability coefficients of d-limonene (a common flavor component in citrus products) in the HDPE liner were three to four orders of magnitude higher than those in glassine. It was also found that the solubility of d-limonene in the glassine liner was substantially lower than in the HDPE liner for the same vapor pressures. Hence equilibrium distribution of the limonene vapor between a product such as a fruit-flavored cereal and the respective liners will result in a much lower limonene concentration within the glassine liner, and "scalping" of the cereal flavor can be assumed to be much more significant in the HDPE liner.

11.4 Pastas

The word "pasta" is traditionally associated with Italy and with wheat semolina, Italy cannot claim to have invented this popular food and semolina is not the original raw material. The Chinese invented pasta that was produced as noodles from rice and legume flours several thousand years B.C. Today pasta consumption is increasing in many countries to the extent that pasta can be classed as a truly international food.

Part of the appeal of pasta products (macaroni, spaghetti, vermicelli, noodles) is that they may be prepared from several raw materials according to countless formulations, and cooked and served in numerous ways to various tastes. Durum wheat semolina is considered to be the best raw material for pasta making because of the functional characteristics of its proteins and its high pigment content. The limited availability and high cost of durum wheat compared with other cereals has led to the use of flours and starches from rice, maize, barley, soft wheat, cassava and potato in pasta formulations in various parts of the world.

The original composition of pasta (water, durum wheat, semolina and egg) has altered considerably to include vitamin supplements, iron salts, powdered vegetables, tomato concentrate, milk protein, other cereal flours, and meat and cheese in filled pasta in order to satisfy the tastes and food habits of different populations. The introduction of modified atmosphere packaging has "enabled certain types of pasta (particularly the fresh and filled varieties) to progress from being a small-scale manufacturing operation to an established position in the food industry where, for example, fresh product is distributed across the USA.

Pasta can be subdivided into two categories. The first, macaroni, has come to represent a generic family of over 140 items in the United States and includes spaghetti, macaroni and vermicelli. This class of product is made from semolina, water, and in most cases, added vitamins and minerals, and contains about 1.5 per cent fat.

The second broad category of pasta is noodles which contain the same ingredients as pasta as well as egg solids at a minimum level of 5.5 per cent. The egg solids are usually in the form of yolks to enhance color and therefore, the lipid content of the noodles is about 4.6 per cent. Noodles may also contain additional protein from sources such as soy flour.

Dried pasta is produced by the reduction of dough moisture content from 30 per cent to about 11 per cent by means of a dehydration process, the length of which is determined by the temperature used. In recent years, the drying of pasta products at temperatures above 60°C has become widely accepted by European pasta manufacturers. Initial benefits realized from high temperature drying were control of bacteria in egg products and shorter drying cycles which permitted more compact drying lines for a given capacity. Currently two different approaches are used by manufacturers to realize the maximum benefits of high temperature drying. In one method high temperature is applied at the initial phase (predrier) of the drying process, with the remainder of the cycle performed at decreasing temperatures. The other method applies high temperatures during the final drying phase after predrying at conventional temperatures. A study which compared the quality characteristics of spaghetti dried by these two high temperature methods with conventional low temperature processes showed that the high temperature processes resulted in superior quality spaghetti.

If the drying stage of pasta, manufacture is not properly controlled; extensive growth of microorganisms such as *Salmonella* spp. and *Staphylococcus aureus* can occur, resulting in a potential hazard to public health. Extensive studies of the microbiology of pasta products have been reported for Canada and the United States; with a few exceptions, all samples met the overall criteria recommended by the International Commission for Microbiological Specifications for Foods.

Two modes have been identified for pasta product failure: moisture gain or loss and color loss. The major mode for pasta failure is moisture gain, the optimum moisture content for pasta storage appearing to be 10–11 per cent. If the pasta moisture content increases to 13–16 per cent, mold growth (which makes the pasta unfit for consumption) and starch recrystallization or retrogradation (which makes the pasta unacceptably tough when cooked) occur. As with other dried products; the optimum moisture content for stability is derived from the maximum "safe" a_w. Complete sorption and desorption isotherms have been obtained for macaroni at 15, 25 and 35°C, for egg noodles at 25 and 45°C and for vermicelli at 27°C.

The traditional packaging material for dried pasta was the paperboard carton, frequently with a plastic window so that the customer could view the contents. Because of the problem of "dusting off" of pasta products and the static attraction of many plastic films for dust, the window was usually made from heavy gauge cellulose acetate as it has a very low static attraction for dust. It has been shown that the creases and end openings on a typical paperboard box increase the overall transmission rate by two times over that of paperboard alone, thereby reducing the shelf life to half under adverse (high temperature and humidity) conditions. In recent years pasta products have been packaged in plastic films such as OPP or coated LDPE.

The production of pasta involves kneading of the dough, followed by extrusion or lamination and then drawing. In the case of special pasta, this latter stage is accompanied by filling, using a cooked meat or vegetable-cheese mixture, thus resulting in a variety of potential microbiological flora. The Barilla style pasta contains salt which lowers its a_w.

Fresh pasta products are not usually subjected to pasteurization but are refrigerated for retail distribution. The microbiological quality of fresh pasta will thus, depend on the quality of the raw materials, the cleanliness and hygiene of the processing environment and equipment, and the handling of the product during production and packing. In addition, the temperature of the product during storage, distribution and retailing is crucial with respect to microbiological quality. However, some companies are now pasteurizing the pasta after packing using hot air and/or microwaves.

These pasteurized fresh pasta products fall into the category of REPFEDs, *i.e.*, refrigerated, processed foods with extended durability.

Chapter 12

Packaging of Dairy Products

12.1 Packaging Materials

Metal Cans

The traditional method for packaging milk powders for consumers uses three piece tinplate cans where the atmospheric air is withdrawn from the powder and replaced with an inert gas such as N_2 prior to seaming the base onto the can. When correctly seamed, the can is essentially impermeable to oxygen, water vapor and light, and can be filled at high speeds. Its mechanical strength facilitates transport and handling, and the reuse possibilities of the empty can may contribute to its popularity in many parts of the world. It is usual for the top of the can to have a lid which can be levered off, and, in order to provide a gas-tight seal under the lid, an aluminum foil diaphragm is sealed to the rim of the can. This is punctured by the consumer immediately prior to use. The use of an easy-open lid incorporating a ring-pull made from aluminum and scored is now quite common; a plastic overclosure is supplied to provide a limited degree of protection once the metal end has been removed.

Laminates

In recent times aluminum foil/plastic laminates (sometimes containing a paper layer as well) have been introduced as a replacement for the tinplate can. The laminates can be formed, filled, gas flushed and sealed on a single machine from reel stock. Gas flushing is achieved by saturating the powder with inert gas but does not include the evacuation step used with cans. The main advantages associated with the laminated type of material are lower material cost and lighter material weight. The disadvantages are that such laminate packs do not have the mechanical strength and durability of rigid containers, and there can be difficulty in obtaining a satisfactory heat seal because of contamination of the heat seal area by powder during filling.

The physical strength of a laminate depends on the material of which it is composed. Aluminum foil 9 μm thick laminated to paper of weight 45 g m^{-2} and LDPE 25 μm thick will give a burst strength of 179 kPa. However, elimination of the paper and the introduction of a PET outer layer will change the physical properties considerably without greatly affecting the chemical properties. For example, 9 μm aluminum foil laminated with 12.5 μm PET and 64 μm LDPE increases the burst strength to 290 kPa.

Form-fill-seal (FFS) machines are usually designed to produce gusseted or ungusseted pouches. As there is less handling of the material with FFS machines during the production sequence, it is possible to omit the paper layer normally incorporated into premade bags. A typical material construction for milk powder on FFS machines is PET (17 μm)–LDPE (9 μm)–foil (9 μm)–LDPE (70 μm).

Fiber Cans

Fiber cans or composites manufactured by spiral winding of paper board strip are obtainable with a wide variety of liners. They can give a similar degree of protection to that obtained from aluminum foil/LDPE/paper bags and a strength similar to a metal can. They have the added advantages of being lighter than metal cans and not corroding, a problem which can occur with metal cans under high humidity conditions. Fiber cans are filled in the same way as metal cans but a disadvantage is that they cannot be cleaned with hot water. A typical specification for fiber cans for use with either whole or skim milk powder is 0.9 mm board and a foil coating of 5 μm with a nitrocellulose lacquer to protect the powder from the aluminum foil. An outer decorative label incorporating a fiber seal material gives increased protection against moisture penetration.

Occluded air trapped in vacuoles within powder particles is not removed during gas flushing of cans or laminate packs. Depending on the volume of occluded air in the powder, the equilibrium residual O_2 content of the pack could exceed the 0.02–0.03 mL g^{-1} normally tolerated. This problem can be alleviated by conditioning the powder under vacuum for 24–48 hours prior to gas flushing to remove occluded air.

Today many types of packaging techniques are employed, for the preservation of dairy products.

Vaccum Packaging

An alternative method of reducing the O_2 content of packaged whole milk powder is by compression of the powder. However, while such a procedure reduces the air content, it results in a significant increase in bulk density and (usually) a decrease in performance as measured by ease of solubility. Vacuum packaging is used to reduce the O_2 content, and it achieves this by evacuation of the interstitial air and some compression as a consequence the vacuumization. The main problem associated with vacuum packaging of milk powder is that of removing air from the package without removing powder fines which could damage the vacuum pump and contaminate the sealing area of the laminate bag. This problem can be avoided by applying the vacuum at a slow rate so that powder particles are not disturbed.

Gas Packing

The technique of gas packing is simple and, provided that the package is airtight, the O_2 content can easily be reduced from the 21 per cent present in air to 1 per cent or less immediately after the operation is finished. However, with spray dried powder this diminution in O_2 content is limited mainly to the atmosphere surrounding the hollow powder particles. Up to 28 days may need to elapse before equilibrium is established between the gases within and around the particles, although most of the change will usually be complete after 7–10 days. During this desorption period the O_2 content within the can may rise to as much as 5 per cent, considerably above the level of 1–1.5 per cent required to ensure optimal keeping quality. When O_2 levels of the order of 1 per cent are needed, there is no alternative in the usual gas packing technique but to store the powder after the initial gas packing long enough for desorption to take place and then to gas pack the cans a second time. This is costly and time consuming, requires considerable storage space, increases the handling which will remove the desorbed O_2 as it is liberated is likely to receive close scrutiny.

One such process was described in 1955, based on the removal of small amounts of O_2 by combination with H_2 in the presence of a palladium or platinum catalyst. The system involved flushing the package containing pellets and product with a combination of N_2 and H_2 gas. Initially the pellets were placed within gas permeable pouches and attached to the inside of the can lid. A comparison of conventional double gas packing with N_2 (with an interval of several days for desorption) with single gas packing with 90 per cent N_2, 10 per cent H_2 and the palladium catalyst for both whole and skim milk spray dried powders showed the new system was more effective in removing oxygen and maintaining an almost oxygen-free atmosphere within the cans. Assuming that the O_2 level within the can could rise to 10 per cent (an unlikely high value), the increase in moisture content caused by the reduction of this O_2 to water would be slightly less than 0.02 per cent. It was concluded that if the process was used on a commercial scale, both whole and skim milk powder would probably be usable after storage for 10 years at normal temperatures.

To avoid the possibility that someone might accidentally ingest the pouch, the American Can Company incorporated the palladium catalyst (in powder form on alumina) into the film construction. The multilayered film (known as Maraflex 7-F) consisted of PET/ adhesive/aluminum foil/ionomer/catalyst/ionomer, an ionomer being chosen because the film between the catalyst and food had to be permeable to H_2 and O_2 as well as form a strong heat seal.

In trials using a gas flush of 8 per cent H_2 and 92 per cent N_2 with 100 g of whole milk powder in pouches, initial O_2 levels of 0.5–2.5 per cent were produced, these dropping to 0.1–0.2 per cent after one day. Two days after packaging O_2, readings were zero and an adequate amount of H_2 remained in excess of the quantity needed to scavenge the O_2. The moisture content increased from 2.42 per cent initially to a maximum of 3.34 per cent after one week and then declined to below the initial level after 2 weeks. Storage conditions were 21°C and 50 per cent RH. Similar results were obtained by other workers. Despite the apparent successes of the system, it has never been adopted commercially.

12.2 The Microorganisms in Raw Milk

In virtually all countries liquid milk for consumption must be pasteurized and cooled before it is packed. The primary purpose of the pasteurization is to destroy any pathogenic microorganisms present in the raw milk to make it safe for human consumption, while at the same time prolonging the shelf life of the milk by destroying other microorganisms and enzymes that might ruin the flavor.

In general, only 90–99 per cent of all microorganisms present in raw milk are destroyed by pasteurization, and pasteurized milk producers depend on refrigerated storage to achieve the required shelf life. Post-pasteurization contamination (PPC) with cold-tolerant Gram-negative rods is the limiting factor affecting the shelf life of commercial pasteurized milks at refrigerate storage temperatures. Inefficient sanitation of milk contact surfaces and contamination from the dairy plant atmosphere are the major causes of PPC, with the packaging machine fillers being considered to be the most important. Upgrading of pasteurized milk handling and packaging systems to ultra- clean standard appears to be an effective method for extending shelf life at refrigeration temperatures. In the absence of PPC, the shelf life of low acid pasteurized milk products depends on the activity of heat-resistant organisms which survive pasteurization, and their level of activity depends on the storage temperature. At 5°C a shelf life of 18–20 days would appear to b realistic for pasteurized milk products processed using an ultra-clean packaging system.

All fluid milk, except for a small quantity of "certified" raw milk, is pasteurized at either 63°C for 30 min (referred to as the Low Temperature Long Time (LTLT) method or sometimes as the holding method) or at 72°C for at least 15 s [referred to as the High Temperature Short Time (HTST) method] and then quickly cooled to 5°C or below. These heat treatments designed to destroy microorganisms that produce disease and to reduce the number of spoilage microorganisms present. They do not sterilize the product.

The disease-producing organisms of chief concern in milk are *Mycobacterium tuberculosis,* a non-spore-forming bacterium that causes tuberculosis and is frequently found in the milk of infected animals; Brucella species that cause brucellosis in animals and man; and *Coxiella burnetti* that causes a febrile disease in humans known as Q fever. The most resistant of these three organisms is *C. burnetti* which is characterized by a $D_{65.6}$ of 30–36 s and a z of 4–5°C. *M. tuberculosis* is characterized by a $D_{65.6}$ of 12–15 s and a z of 4–5°C, and Brucella species by a $D_{65.6}$ of 6–12 s and a z of 4–5°C. It is left as an exercise for the interested reader to calculate the number of decimal reductions achieved for each of the above microorganisms by the LTLT and HTST pasteurization processes.

Now let us have look at the factors which affect the shelf life of pasteurized dairy products.

One of the most critical factors affecting the shelf life of pasteurized dairy products is the temperature of storage. Attention has been focussed on ways of predicting the effect of temperature on the growth of bacteria in foods. The "square root" equation has been found to provide a good description of the growth of a number of psychotrophic bacteria in dairy products.

The shelf life of pasteurized milk has been found to be determined mainly by the level of contamination with Gram-negative psychotrophic bacteria; the effect of temperature on growth of contaminants could be accurately determined by the "square root" plot but the conceptual minimum temperature for growth (T_o) varied. The variation was related to the quality of the pasteurized milk. The microflora of pasteurized milk varied significantly with storage temperature. Thus, while spoilage at refrigeration temperatures was mainly due to the growth of Pseudomonas spp., Enterobacteriaceae and Gram-positive bacteria assumed greater importance in the spoilage of milks stored at temperatures above 10°C. It has been suggested that 15°C may be a useful preincubation temperature for predicting the shelf life of pasteurized milk samples.

Effect of Light

During processing, distribution, storage and marketing, milk may be exposed to natural and artificial light. Flavor changes as well as loss of vitamins and other nutritional components are attributed to chemical reactions induced in the milk by light with wavelengths below 550 nm. This problem is quite widespread, a study in 1984 finding that 40 per cent of the milk samples in plastic purchased from stores showed light-induced off-flavors. Responses from taste panels have indicated that in the early stages of oxidation milk loses its naturally fresh flavor and becomes quite flat in taste without being objectionable as in the more advanced stages of oxidation.

Riboflavin (vitamin B_2) plays a central role as it is not only destroyed itself by light but, in addition, catalyzes the development of oxidized flavor and ascorbic acid oxidation by generating excited state (singlet) oxygen. Because of its absorption by riboflavin, light of wavelengths of 350–550 nm is the most damaging, the maximum damage occurring at about 450 nm. These wavelengths are contained in the emission spectra of white fluorescent tubes.

Riboflavin destruction by light has been shown to be greater in skim milk than whole milk, since light of 400–500 nm wavelengths can penetrate 40–50 per cent deeper into skim milk than whole milk. Destruction of nutrients by light could present legal problems in connection with nutrient standards and/or labels, since the milk after storage in supermarket cabinets may not meet regulatory requirements.

In a study on the relative destruction of vitamin A and riboflavin in skim milk when exposed to fluorescent light at an intensity of 2000 lumens m^{-2} for 24 hours, more than 75 per cent of the added vitamin A was destroyed in glass, clear polycarbonate and polyethylene containers. Paperboard containers provided the most protection, while gold-tinted polycarbonate which blocks light of 400–480 nm provided the second best protection.

Plastic sleeves made from poly(methyl methacrylate) with a yellow-green tint are available for installation over fluorescent tubes in display cases to shield products from light. Such tubes have 92 per cent transmission in the region from 800 to 440 nm, rapidly decreasing to zero per cent transmission at 385 nm and below. However, these shields would provide little protection from light in the damaging 400–550 nm wavelength region.

The extent of nutrient loss in a supermarket depends on the proximity of the containers to the light source, the number and wattage of light bulbs in the display case, the size of the exposed surface area, and the length of exposure. The detrimental effects of fluorescent light on milk could be alleviated by selection of packaging materials to minimize transmission of light; reduction of light intensities in display cases to 500 lumens m^{-2}; use of yellow or yellow-green lamps or filters in display cases; and responsible rotation at retail outlets to limit prolonged detrimental exposure to light.

It is clear that exposure of milk to light in the 400 to 550 nm wavelength region can result in the development of off-flavors and destruction of nutrients. It has been recommended that the packaging materials used for milk should not transmit more than 8 per cent of incident light at 500 nm wavelength and not more than 2 per cent at 400 nm. Exposure to direct sunlight should be avoided under all circumstances as this also tends to increase the temperature of the milk which accelerates microbial spoilage.

Effect of Oxygen

Oxygen plays an important role in the light-induced development of off-flavors in milk. Pasteurized milk at filling is generally saturated with O_2 (about 8 ppm), but if no additional O_2 can gain access, its content falls and the rate of the adverse reactions slows or stops. However, additional O_2 from a headspace or entering through a permeable container will maintain the O_2 content and keep the rate of oxidative reactions high. Oxygen has been shown to be limited in glass bottles because, unlike cartons and polyethylene bottles, they are impermeable and O_2 used up in oxidative reactions cannot easily be replenished.

Packaging Materials

Milk for retail sale is packaged in glass, coated paper and plastic containers of various compositions and constructions. Both single service and multiuse (refillable) containers are used, although the trend in recent years has been towards single serve packaging.

It has been demonstrated that the packaging material is a key to the protection of the flavor and nutritional qualities of fluid market milk. Numerous studies have shown that riboflavin is destroyed by light of the same wavelength as that which produces light-induced off-flavor.

Although the use of brown glass in milk bottles to reduce light damage was proposed more than seventy years ago it has only been adopted commercially in a few countries (*e.g.*, Sweden). The total amount of light passing through the container wall depends on the material from which the container is made and also on the color either incorporated into the material (*e.g.*, the pigment in the glass) or used in printing it. The color determines the wavelength of the light reaching the milk. Ruby glass is a most efficient barrier and amber is also very effective, but cost and marketing considerations militate against both.

Unpigmented polyethylene milk bottles in the 350 to 800 nm spectral region have been found to transmit 58 to 79 per cent of the incident light. Light transmission was reduced by pigmenting with titanium dioxide (1.6 per cent), the bottle being opaque below 390 nm. Use of a colorant with the titanium dioxide pigment further reduces transmission of light with wavelengths less than 600 nm. The unprinted area of a paperboard carton had less than 1.5 per cent transmission below 550 nm and was opaque to wavelengths below 430 nm.

In a study on light induced quality deterioration of milk four common milk packaging materials were used: clear polyethylene pouch; coextruded laminate polyethylene pouch (outer white layer and an inner black pigmented layer); standard paperboard carton, and returnable plastic jug. Off-flavor was detected in all containers except the laminate pouch.

A more recent study compared milk packaged in white polyethylene-coated cartons overprinted with blue, and polyethylene bottles. Cartoned milk was disliked by a flavor panel after about 17.5 h exposure to 4000 lux at 7°C and milk in the polyethylene bottles after 9 h exposure. The dissolved O_2 concentration dropped considerably in the bottled milk exposed to light, but only marginally in the cartons, light induced flavor development depending on O_2 availability. Because glass bottles are essentially impermeable, O_2 used up in oxidative reactions cannot easily be replenished and is likely to be a limiting factor.

From the quite extensive results now available in the literature, it would appear that the most efficient light barrier is the paperboard carton. This was not true of the early waxed cartons when wax penetrated the board, rendering it translucent. However, the modern polyethylene-coated board is very efficient and can be made even more so by overall printing, particularly in reds, yellows, blues, greens, browns and black in ascending order of efficiency.

Reuse of milk containers poses several problems to the consumer and the industry in that the industry has no way of controlling what a consumer may place in the container before it is returned for refilling, nor does the consumer have any way of knowing how a container may have been misused before being returned to the milk plant.

In a study where glass, polyethylene and polycarbonate multiuse milk containers were treated with 29 common household chemicals to simulate consumer abuse, glass was found to be the easiest to clean and was most resistant to retention of treatment contaminants. Neither the polyethylene nor the polycarbonate multiuse milk container appeared to comply with the provisions of the Grade "A" Pasteurized Milk Ordinance Recommendations of the US Public Health Service.

12.3 The Packaging of Fresh Cream and Fermented Milks

Cream may be defined as that part of milk rich in fat which has bee separated by skimming or mechanical separation. Cream may be free flowing and easily pourable, highly viscous and pourable only with difficulty, or spoonable but non-pourable. The method of production, the milk fat content and the system of treatment can have a marked effect on the physical characteristics of the cream.

The fat content of cream may range from 10 per cent (half-cream) to 80 per cent plus (plastic cream); cream for butter manufacture would normally contain approximately 40 per cent fat. The United Nations FAO/WHO have suggested the following standards for Market Cream:

1. Pasteurized, sterilized and UHT treated cream > 18 per cent milk fat.
2. Half-cream, 10–18 per cent milk fat.
3. Whipping cream > 28 per cent milk fat.
4. Heavy whipping cream > 35 per cent milk fat.
5. Double cream > 45 per cent milk fat.

For many years it was customary to package fresh cream in waxed paper cartons with press-in lids of the same material. However, on storage under refrigerated conditions the filled cartons showed a marked tendency to absorb moisture and become deformed.

The waxed paperboard cartons were eventually superseded by plastic tubs (usually made from high-impact PS) of similar shape and sizes. These were initially covered with crimped-on skirted caps of aluminum foil, and later by aluminum foil heat sealed to the rim of the container. In addition, rectangular gable-topped waxed or polyethyle-coated paperboard packages are used, sometimes incorporating aluminum foil as a barrier in the laminate. The traditional glass bottle with a foil cap is still used in some parts of the world, but is increasingly being replaced by blow-molded polyethylene containers of similar shape, sealed with a close fitting plastic cap.

Fermented milks are products prepared from milks (whole, partially or fully skimmed milk, concentrated milk, or milk reconstituted from partially or fully skimmed dried milk) homogenized or not, pasteurized or sterilized and fermented by means of specific microorganisms.

Yoghurt, the most important of the fermented milk products, is a coagulated milk product obtained by lactic acid fermentation through the action of *Lactobacillus bulgaricus* and *Streptococcus thermophilus*. It may contain added fruit or fruit flavors as well as carbohydrate sweetening matter. Kefir has an alcohol content of 0.5–2 per cent and koumiss of 2–3 per cent, and both contain considerable quantities of CO_2 which is formed by the heterofermentative, aroma-forming lactic acid bacteria.

Table 12.1: Composition of Fermented Milks

Dry matter	14–18 per cent
Protein	4–61 per cent
Fat	0.1–10 per cent
Lactose	2–3 per cent
Lactic acid	0.6–1.3 percent
Fruits and carbohydrate sweeteners	5–25 per cent
pH	3.8–4.6

Fermented milks can be classified into three consistency groups: set (firm), stirred (pasty) and liquid. The coagulum of firm products can be destroyed and whey expelled as a result of mechanical shear.

Although this effect can be reduced by using packages which are conical in shape with the larger diameter forming the base, containers of this shape are unable to be stacked.

The protective effect of five different types of containers for packaging solid unflavored yogurt stored at 8°C in the dark and under 2000 lux for 12 hours per day has been evaluated. The results are summarized in Table 12.2, with the rankings being based on the results of sensory analyses and the determinations of peroxide value, vitamins A and B_2, and instrumental color measurements. The transparent, brown-pigmented glass jar was found to offer superior protection against light and O_2. The authors recommended that at a storage temperature of 8°C, the shelf life of yoghurt package nonaseptically in the brown glass container was 16–18 days.

Table 12.2: Summary of Protective Effects of Different Packaging Materials for Yoghurt

Materials	Protection Against		Rank in Order of Decr-easing Total Protection
	Light	*Oxygen*	
Transparent brown glass	Good	Perfect	1
Unpigmented glass	Moderate	Perfect	4
Transparent brown PS	Good	Moderate	3
Unpigmented PS	Bad	Moderate	5
Paperboard and PS	Excellent	Bad	2

The level of CO_2 in fermented milk products such as kefir contributes to the flavor and to the growth of the starter organisms. Thus, the influence of the packaging material on the CO_2 concentration in the product is clearly important. A study comparing regular LDPE-coated paperboard cartons with cartons made from a special board with a double thickness of LDPE on the inside and an aluminum foil lined board found that the CO_2 content dropped to 25 per cent or less of its original concentration in the first two types of containers after 13 days of storage, while it was maintained in the foil-lined cartons. It is also important to control the yeast content in kefir, as too high a content gives a pronounced yeasty flavor which is undesirable. After 13 days storage, the yeast content in the foil-lined carton has increased by only 60 per cent compared to a more than 6-fold increase in the LDPE-lined cartons.

12.4 The Packaging of Cheese

Cheese is the generic name for a group of fermented milk-based food products produced in at least 500 varieties throughout the world. Although some soft cheese varieties are consumed fresh, *i.e.*, without a ripening period, production of the vast majority of cheese varieties can be sub-divided into two well-defined phases; manufacture and ripening. Despite differences in detail in the manufacturing processes used for individual varieties, the conversion of milk into cheese generally comprises four stages.

Coagulation

Physicochemical changes in the casein micelles due to the action of proteolytic enzymes and/or lactic acid lead to the formation of a protein network called coagulum or gel.

Drainage

Separation of the whey, after mechanical rupture of the coagulum, by molding and in certain cases by pressure, to obtain a curd.

Salting

Incorporation of salt by deposition on the surface or within the body of the cheese, or by immersion in brine.

Ripening

Biochemical changes in the constituents of the curd brought about by the action of enzymes, mostly of microbial origin.

Cheese manufacture can be viewed as essentially a dehydration process in which the fat and casein in milk are concentrated between six- and twelvefold, depending on the variety. The degree of hydration is regulated by the extent and combination of the three steps listed above, in addition to the chemical composition of the milk. In turn, the level of moisture in the cheese, the salt content and the cheese microflora regulate the biochemical changes that occur during ripening and hence determine the flavor, aroma and texture of the finished cheese. Although the nature and quality of the finished cheese are determined in very large measure by the manufacturing steps, it is during the ripening phase that the characteristic flavor and texture of the individual cheese varieties develop.

Although some soft-cheeses are consumed fresh, most cheese varieties are not ready for consumption at the end of manufacture but undergo a period of ripening (also referred to as curing or maturation) which varies from about 4 weeks to more than 2 years. The duration of ripening is generally inversely related to the moisture content of the cheese, although many varieties may be consumed at any of several stages of maturity depending on the flavor preferences of consumers.

A great variety of microbial species are involved in the ripening of cheese, the total population generally exceeding 10^9 organisms per gram. The principal bacterial groups involved in ripening are the lactic acid Streptococci; Leuconostoc; Lactobacilli and Propionibacteria. Micrococci and Corynebacteria species may also be involved; they are aerobic and salt tolerant and thus, grow especially on the surface of cheeses.

Yeasts are widely distributed in nature and are found in raw milk and some cheeses, the basic flora in the majority of cheeses being species of the genus *Kluyveromyces*. Yeasts produce enzymes capable of degrading the constituents of the curd and so can contribute to modifying the texture of the cheese and to the development of flavor and aroma. They are capable of converting lactose into CO_2 and may also take part in lipid degradation. The role of yeasts in the ripening of certain cheeses may be complex and significant.

Among the fungal flora found in or on mold-ripened cheeses, species of the genus *Penicillium* are of particular importance. *Penicillium camemberti* is the original mold of Camembert and Brie and has a single habitat–the surface of a few cheeses and the environment of cheese factories. By contrast, *Penicillium roqueforti,* a microaerophilic mold, is widely distributed and is the mold of mold-ripened cheese; it grows very well at quite low O_2 levels (5 per cent). *Geotrichum candidum* is present on certain soft cheeses where it forms a characteristic grayish-white crust on the surface. The role played by molds in ripening is a major one for soft and mold-ripened cheeses.

The oxygen requirements of microorganisms are known to be variable. Some are strictly anaerobic and will grow only in the interior of the cheese, *e.g.,* Clostridia and Propionibacteria. Other microorganisms are microaerophilic and grow best in an environment with a low oxygen content, *e.g.,* lactic acid bacteria, especially Lactobacilli. Others are strictly aerobic and cannot grow actively except on the surface of the cheese, *e.g.*, molds, yeasts, micrococci and coryneform bacteria. Cheese ripened by the action of these microorganisms will necessarily be of small size. One exception is cheeses ripened by internal mold (*Penicillium roqueforti*), but in this case a certain aeration of the cheese is ensured by piercing, and the mold is able to grow in a relatively low O_2 atmosphere.

The composition of the atmosphere may also directly affect ripening. Thus ammonia, which is present in the ripening cellars of washed-rind soft cheeses, helps to neutralize the surface layer of the cheese. In this way it favors the growth and activity of the surface bacterial flora, *i.e.*, micrococci and coryneform bacteria. Very little is known about how the gaseous composition of the atmosphere in the factory, the packaging or the interior of the cheese influences the microbial, enzymic and chemical phenomena of ripening. Despite this, it is one of the important methods of achieving control over the process.

Water activity in cheese is determined by two basic parameters: water content and salt content. During ripening the a_w is not stable but changes constantly, reducing until the a_w of the surface is in equilibrium with the surrounding atmosphere. There is a gradient of a_w from the moister central zone of the cheese to the drier zone near the surface. Overall, the a_w of the whole cheese is the resultant of a set of complex physical phenomena (drainage; evaporation of water; diffusion of salt); chemical phenomena (interaction between the substrate and sodium chloride); and biochemical phenomena (proteolysis; lipolysis; glycolysis).

The salt concentration in the aqueous phase of most cheeses lies between 4 and 5 per cent, reaching 6–8 per cent in blue veined cheeses. These concentrations have a selective action at the microbial level and an inhibiting effect on enzyme activity. The selective action of salt on the growth of molds on the surface of soft cheese is well known: a reasonable degree of salting limits or even prevents the growth of *G. candidum*, without noticeably affecting that of *Penicillium*.

Fermentation of lactose to lactic acid during cheese manufacture reduces the pH of cheese curd to values that inhibit the growth of many pathogenic bacteria. The formation of lactic acid during manufacture and the metabolism of residual lactose during the initial stages of ripening reduce the pH of cheese to around 5 depending on variety. During ripening, the pH of the cheese rises due to the formation of alkaline N-containing compounds and or the catabolism of lactic acid.

The influence of pH on microbial growth and enzyme activity is particularly decisive. Only lactic acid bacteria, yeasts and molds can grow at pH values below 5. Enzymes are also very sensitive to variations in pH, most microbial proteases having greatest activity in the pH range 5.0–7.5, and lipases in the range 7.5–9.0. Curd at the end of drainage has a pH of less than 5.5 and is the site of active lactic acid fermentation. This acidic character of the curd is a necessity in that it slows down enzyme action and impedes the development of a harmful bacterial flora.

Soft cheese curd has a pH below 5, being around 4.5 for Camembert. Such a pH does not allow sufficient protease activity. Some neutralization of the curd is therefore necessary and this occurs either by the growth of molds and yeasts using lactic acid as a source of carbon in the case of semisoft surface molded cheeses of the Camembert type, or by the combined action of yeasts and the ammoniacal atmosphere of ripening rooms in the case of washed-rind soft cheeses. The pH of Camembert increases from 4.5 to over 6.0 in the interior and over 7.0 on the surface by the end of ripening (about 30 days).

Hard cheeses have pH values around 5.2–5.3 and neutralization of the cheese is undesirable. The pH tends to remain at about this level during ripening, partly because of the high degree of mineralization of the cheese which gives it a high buffering capacity.

The two key parameters contributing to the stability of cheeses are pH and a_w. However, neither of these parameters are low enough to ensure complete stabilization of the product, with the result that cheeses as a class lie between perishable foods on the one side and intermediate moisture foods on the other. While the packaging will have no influence on the pH of the cheese, the a_w of the surface (and ultimately the interior) of the cheese will be affected by the water vapor permeability of the packaging material.

Two other key factors which must be considered in the packaging of cheese (and indeed in the packaging of all dairy products) are the effect of light and oxygen. As discussed earlier in this chapter, light initiates the oxidation of fats, even at temperatures found in refrigerated display cabinets, giving rise in the case of unripened cheeses to off-flavors which have been described as "cardboardy" or "metallic". The oxidation reactions initiated by light may continue even if the cheese is subsequently protected from light, and any metallic ions contained in the packaging material will catalyze the reactions. In addition, the ingress of molecular oxygen through the packaging film is undesirable as it will contribute to the oxidation of fats and the growth of undesirable microorganisms.

Several attempts have been made to develop a classification system for cheese varieties and these have been reviewed by Fox. There is no one classification system applicable to all cheeses, but for the purposes of providing a framework to discuss packaging requirements, the following scheme is appropriate and will be used here: very hard and hard; semisoft and soft; fresh; and processed.

The Very Hard class of cheese (sometimes referred to as grating grade) is ripened by bacteria and is characterized by a moisture content (on a fat free basis) of <51 per cent, *e.g.*, Parmesan 42 per cent; Romano 31 per cent; Mozzarella 45 per cent. The Hard class of cheese is also ripened by bacteria and is characterized by a moisture content of 49–63 per cent, the range 49–56 per cent being categorized as hard, and the range 54–63 per cent being categorized as semi-hard. Cheeses in this class include Cheddar (still the cheese produced in the greatest quantities worldwide), Edam, Gouda, Cheshire, Gloucester, Derby and Leicester, as well as those with eyes such as Emmental and Gruyere. Also included in this class are Provolone, Mozzarella and Kasseri.

Table 12.3: Recommendations Regarding the Permeability of Packaging Materials[a]

Cheese Category	Water Vapor Transmission Rate $g\ m^{-2}\ day^{-1}$ at 25°C/75%RH	O_2 Transmission Rate $cm^3\ m^{-2}\ day^{-1}\ atm^{-1}$ at 25°C/75%RH
Consumer-packaged	30	12,900
Ripening film for whole cheese	2.5	30
Gas or vacuum packaged	4.0	170

[a] From British Standard BS 4855:1973.

In Cheddar and many of the other varieties in this class, lactose is metabolized to L-lactic acid which is then isomerized to D-lactate by non-starter lactic acid bacteria (NSLAB). In Swiss cheese, L-lactate and, to a lesser extent, D-lactate are converted to propionate, acetate and CO_2 when the growth of propionic acid bacteria is initiated on transfer of the cheese to the hot room.

Now let us study about the role of oxygen in the packaging of cheese.

The fate of oxygen in the cheese is still not completely understood. What is certain is that it is the gas present in the space between the packaging material and the cheese (either produced as a product of enzymic action in the cheese, left in the package after sealing, or diffusing through the packaging material) which determines whether or not microbial growth will occur on the surface of the cheese.

The secondary microflora of Cheddar cheese consists of lactic acid bacteria that gain access accidentally, usually as a result of contamination following pasteurization of the milk. These adventitious organisms may grow in the highly selective cheese environment such that certain non-starter lactic acid, bacteria (NSLAB) often attain high densities ($> 10^7$ colony-forming units per gram of cheese) which are maintained for months. Thus, adventitious bacteria are the dominant flora in cheese throughout much of the long ripening period.

In the presence of O_2 certain NSLAB convert lactate to acetate, CO_2 and water. The extent to which lactate oxidation occurs is dependent on the type of NSLAB present and their concentration in the cheese, which depends largely on the storage temperature and time. Also important are the surface area : volume ratio of the cheese and the O_2 permeability of the packaging material. The latter will vary with temperature and may vary with relative humidity, depending on the chemical nature of the material.

Packaging film for Cheddar cheese must be sufficiently impermeable to O_2 to prevent fat oxidation and mold growth. Another factor in mold growth may be the presence of O_2–consuming NSLAB in the cheese since these organisms will compete for O_2 permeating through the film, thereby reducing its concentration under the film. When there is a high density of NSLAB possessing the lactate oxidation system then the cheese will act as an O_2 sink and mold growth may be suppressed.

Since Cheddar cheese contains approximately 1.5 per cent w/w lactate, the, theoretical upper limit for the acetate concentration arising from lactate is about 1 per cent. In practice this level is never reached, mainly because of O_2 limitation. The major free fatty acid in Cheddar cheese is usually acetate and, it is considered by some to contribute to the flavor of the cheese, although concentrations greater than about 0.07 per cent w/w in young cheese give rise to off-flavors. The potential acetate production by NSLAB resulting from O_2 penetration into the cheese is another reason to use packaging materials with low O_2 permeabilities.

Data relating to the minimum partial pressure of O_2 necessary for the development of molds is scarce. In the absence of CO_2, molds will grow at O_2 partial pressures below 1 mm Hg. In the presence of CO_2 at partial pressures: of 65–110 mm Hg (8.5–15 per cent), microbial growth is inhibited up to an O_2 partial pressure of 15–20 mm Hg.

In a relatively early study to determine the maximum O_2 permeability requirements of plastic films that would ensure mold-free Provolone cheese (an Italian hard cheese chosen for the study because of its unusual sensitivity to mold growth), it was concluded that the O_2 permeability of the packaging material should not exceed 0.5 cm^3 100 in^{-3} day^{-1} atm^{-1} when stored under the customary conditions of 5°C and 76–85 per cent RH. Such a film was a laminate of LDPE (3.08×10^{-3} cm thick) and PET (1.27×10^{-3} thick).

Table 12.4: Maximum O_2 Transmission Rate[a] of Packaging Material for Rindless Cheese[b]

	O_2 Transmission Rate	
	Non-shrink Film	Shrink Film
Mass 5 kg and heavier		
Cheddar type cheese	50	75
Gouda, Swiss and Edam type cheese	170	220
Mass up to 5 kg		
All types	120	300

[a] cm^3 m^{-2} day^{-1} measured at 23°C and 93 per cent RH

[b] From Australian Standard AS 2183-1978.

Table 12.5: Composition of the Films Proposed for the Packaging of Rindless Cheeses

	Outside			Inside
Material	First Layer	Second Layer	Third Layer	
1	Copolymer of ethylene and vinyl acetate	copolymer of vinylidene chloride and vinyl chloride	Irradiated copolymer of ethylene and vinyl acetate	
	Thickness: 30 microns	Thickness: 8 microns	Thickness: 12 microns	
2	Copolymer of ethylene and vinyl acetate	Copolymer of vinylidene chloride and vinyl chloride	Irradiated copolymer of ethylene and vinyl acetate	
	Thickness: 15 microns	Thickness: 25 microns	Thickness: 12 microns	
3	Nylon 6 (m.pt. 208°C)	Surlyn ionomer (sodium form)		
	Thickness: 30 microns	Thickness: 30 microns		
4	Mixture of 30% nylon 6 and 70% Surlyn ionomer (zinc form)	Copolymer of ethylene and 4.5 per cent vinyl acetate		
	Thickness: 49 microns	Thickness: 89 microns		
5	Surlyn ionomer (sodium form)	Copolymer of ethylene and 3.5 per cent vinyl acetate colored pale yellow by an inorganic pigment		
	Thickness: 29 microns	Thickness: 106 microns		
6	Copolymer of ethylene and 3.5% vinyl acetate colored pale yellow by an inorganic pigment	Surlyn ionomer (sodium form)		
	Thickness: 26.5 microns	Thickness: 108 microns		

Other published figures for the O_2 permeability requirements of packaging materials for hard cheese include those in Australian Standard AS 2183–1978. Without any indication as to how these values were arrived at, it is difficult to comment in detail. Also, without any mention of the degree of shrink expected or allowed for with the shrink films, it is not possible to assess whether or not the O_2 permeability requirements are identical for cheeses packaged in shrink and non-shrink film. The standard also contains a minimum value for the CO_2 transmission rate for Swiss-type cheeses of 680 cm^3 m^{-2} day^{-1} at 23°C and 93 per cent RH.

Figures for the O_2 transmission rates of cheese bags used in 1986 to package 20 kg blocks of Cheddar cheese in New Zealand have been reported to be 27 cm^3 m^{-2} day^{-1} at 10°C and 100 per cent RH.

The utility of such published data is debatable, but it is likely that problems of mold growth on hard cheeses packaged in the films in common use today are a function of hygienic conditions in the packing room, the degree of vacuum inside the package and the integrity of the heat seal, rather than the O_2 transmission rate of the packaging material used.

Table 12.6: Permeability[a] of the Films Proposed for the Packaging of Rindless Cheeses

Material	To Oxygen cm^3 m^{-2} day^{-1} atm^{-1}		To Carbon Dioxide cm^3 m^{-2} day^{-1} atm^{-1}		To Water Vapor g m^{-2} day^{-1} at 38°C
	0% RH	80% RH	0% RH	80% RH	90% RH
1	155	155	765	765	17.0
2	250	250	1500	1500	29.5
3	25	125	85	455	6.2
4	600	595	2120	2325	4.2
5	955	865	3685	3295	2.3
6	440	640	1470	1565	4.4

[a] According to DIN 53380 Standard.

The development of the technique of ripening under film (also known as rindless ripening) was motivated by various factors. One was the need to roduce blocks better suited to a high degree of mechanization. Another was to increase cost-effectiveness by

replacing earlier techniques of coating, thus reducing handling
during ripening and losses due to drying-out. The first techniques
(which appeared around 1930 and were used for cheddar) involve
coating the blocks with mineral oil. In 1950 J.B. Stine of Kraft Foods
Company filed two patents concerning the use of a plastic packaging
material for blocks of Swiss cheese; these patents have since passed
into the public domain. A further factor was the development of
prepacked, self-service, consumer portions of cheese for or by
supermarkets, where the losses resulting from removal of the rind
were considered unacceptable. This led to the idea of rindless cheese.
There are three variables which may be used to adjust the process of
ripening rindless cheese: the permeability of the packaging material;
the ripening temperature, and the ripening time.

Chapter 13

Packaging of Fruits and Vegetables

13.1 Introduction

Growth, maturation and senescence are the three important phases through which fruits and vegetables pass, the first two terms often being referred to as fruit development. Ripening (a term reserved for fruit) generally occurs during the later stages of maturation and is considered the beginning of senescence, a term defined as the period when anabolic or synthetic biochemical processes give way to catabolic or degradative processes leading to aging and final death of the tissue.

Now let us study about respiration in fruits and vegetables.

Respiration involves the oxidation of energy-rich organic substrates normally present in cells such as starch, sugars and organic acids, to simpler molecules (CO_2 and H_2O) with the concurrent production of energy and other molecules which can be used by the cell for synthetic reactions. The greatest yield of energy is obtained when the process takes place in the presence of molecular oxygen. Respiration is then said to be aerobic. If hexose sugar is used as the substrate; the overall equation can be written as follows:

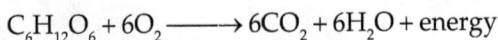

$$C_6H_{12}O_6 + 6O_2 \longrightarrow 6CO_2 + 6H_2O + energy$$

This transformation actually takes place in a large number of individual stages with the participation of many different enzyme systems. The water produced remains within the tissue but the CO_2 escapes and accounts for part of the weight loss of harvested fruits and vegetables, a range of 3–5 per cent of weight loss having been ascribed to respiration. When one mole of hexose sugar is oxidized, 36 moles of ATP (each possessing 32 kJ of useful energy) are formed. This represents about 40 per cent of the total free energy change, the remainder being dissipated as heat. Rapid removal of this heat is usually desirable and it is important that the packaging assists rather than impedes this process.

The rate of respiration is often a good index to the storage life of horticultural products: the higher the rate, the shorter the life, and the lower the rate the longer the life.

From measurements of CO_2 and O_2 it is possible to evaluate the nature of the respiratory process. The ratio of the volume of CO_2 released to the volume of O_2 absorbed in respiration is termed the respiratory quotient (RQ). Values of RQ range from 0.7 to 1.3 for aerobic respiration, depending on the substrate being oxidized for carbohydrates, RQ = 1; for lipids RQ < 1; for organic acids RQ > 1.

Anaerobic respiration (sometimes called fermentation) involves the incomplete oxidation of compounds in the absence of O_2 and the accumulation of ethanol, acetaldehyde and CO_2. Much lower amounts of energy (2 moles of ATP) and CO_2 are produced from 1 mole of hexose sugar than that produced under aerobic conditions. As well, very little heat energy (approximately 5 per cent) is produced for a given amount of carbohydrate oxidation in anaerobic respiration compared to aerobic respiration. The oxygen concentration at which a shift from aerobic to anaerobic respiration occurs varies among tissues and is known as the extinction point. Very high RQ values (> 1.3) usually indicate anaerobic respiration.

Variations in the rate of respiration occur during organ development; as fruit increase in size, the total amount of CO_2 emitted

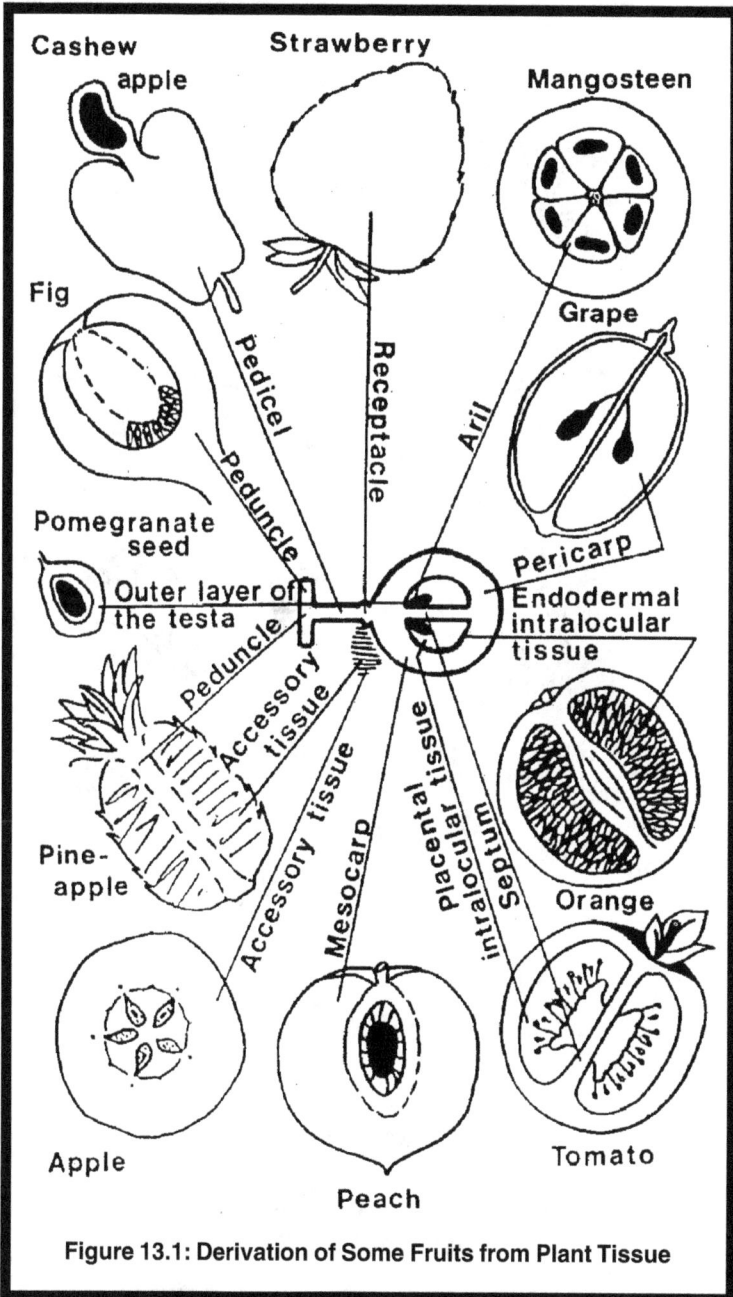

Figure 13.1: Derivation of Some Fruits from Plant Tissue

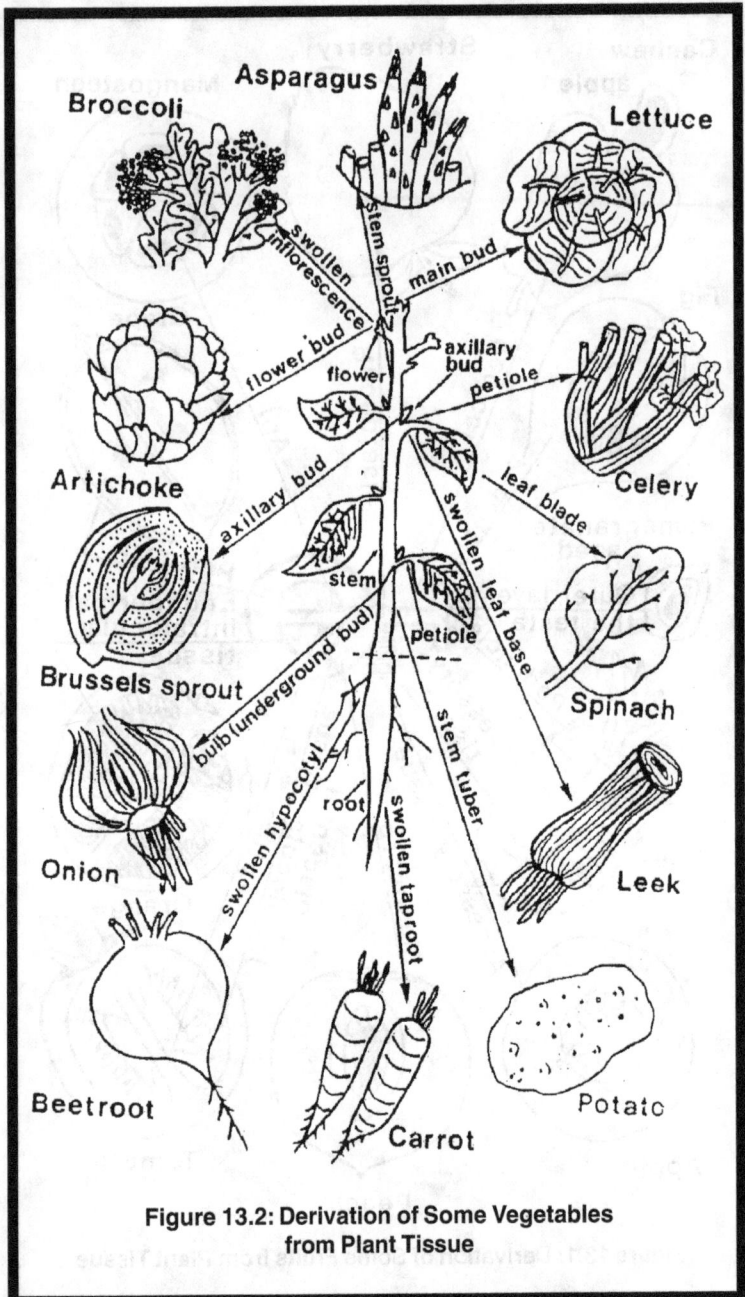

Figure 13.2: Derivation of Some Vegetables from Plant Tissue

increases although the respiration rate calculated on a per unit weight basis decreases continually. Three different routes are available for the exchange of gases between horticultural commodities and their surrounding atmospheres: lenticels and stomata, the cuticle, and the stem scars or floral (calyx) end. In leaves, the resistance of the epidermal layer to gas diffusion is mainly regulated through the stomatal aperture. A majority of studies on gas exchange in plant tissues indicate that the skin represents the main significant barrier to diffusion. Clearly the rate at which the three gases O_2, CO_2 and C_2H_4 diffuse through the tissue will have a significant effect on the rate of respiration.

Table 13.1: Classification of Vegetables According to Respiration Intensity

Class	Respiration Intensity at 10°C mg CO_2 kg^{-1} h^{-1}	Commodities
Very Low	Below 10	Onion
Low	10–20	Cabbage, cucumber, melon, tomato, turnip
Moderate	20–40	Carrot, celery, gherkin, leek, pepper, rhubarb
High	40–70	Asparagus (blanched), eggplant, fennel, lettuce, radish
Very high	70–100	Bean, Brussel sprouts, mushroom, savoy cabbage, spinach
Extremely high	Above 100	Broccoli, pea, sweet corn

13.2 Packaging of Fresh Horticultural Produce

As stated earlier, a simple consideration of equation would suggest that if the CO_2 in the atmosphere were augmented (or the O_2 decreased), the respiration rate and storage life would be extended. The first scientific study on the effect of modified atmospheres on the shelf life of horticultural products was carried out by the French chemist Berard in the 1800s; he found that fruit in an anaerobic atmosphere failed to ripen. Research into the influence of O_2 and CO_2 levels on the chill storage life of apples, pears and berries was carried out in Cambridge, England in the 1920s by Kidd and West.

Their results demonstrated that low O_2 and moderately high CO_2 levels around the fruit reduced the rate of deteriorative reactions.

Table 13.2: Classification of Horticultural Produce According to the Plant Organ Used

Class	Commodities (Examples)
Root vegetables	Carrot, celery, garlic, horseradish, onion, parsnip, radish, turnip
Tubers	Potato, yam, Jerusalem artichoke
Leaf and stem vegetables	Brussel sprouts, cabbage, celery, chicory, Chinese cabbage, cress, green onion, kale, lettuce, spinach
Flower vegetables	Artichoke, broccoli, cauliflower
Immature fruit vegetables	Bean, cucumber, gherkin, okra, pea, pepper, squash, sweet corn
Mature fruit vegetables	Melon, tomato
Reproductive organs	Most fruits

Historically, atmospheres surrounding fruits and vegetables have been altered in controlled atmosphere (CA) storage facilities where the levels of gases are continually monitored and adjusted to maintain the optimal concentration. Because CA storage is capital intensive and expensive to operate, it is more appropriate for those fruits and vegetables that are amenable to long-term storage such as apples, kiwifruit, pears and cabbages. Modified atmosphere (MA) storage typically involves some initial modification of atmospheric conditions which change further with time as a result of the commodity's respiration and the surrounding physical environment.

The recommended concentrations of O_2 and CO_2 for CA and MA storage of fruits and vegetables can be found in the published literature and will not be tabulated here. A very effective way of plotting this data has been presented with CO_2 concentration as the ordinate and O_2 concentration as the abcissa.

Since the 1960s attempts have been made to create and maintain modified atmospheres within plastic polymeric films. The availability of absorbers and adsorbers of O_2, CO_2, C_2H_4 and water provides additional foods for the packaging technologist to use to

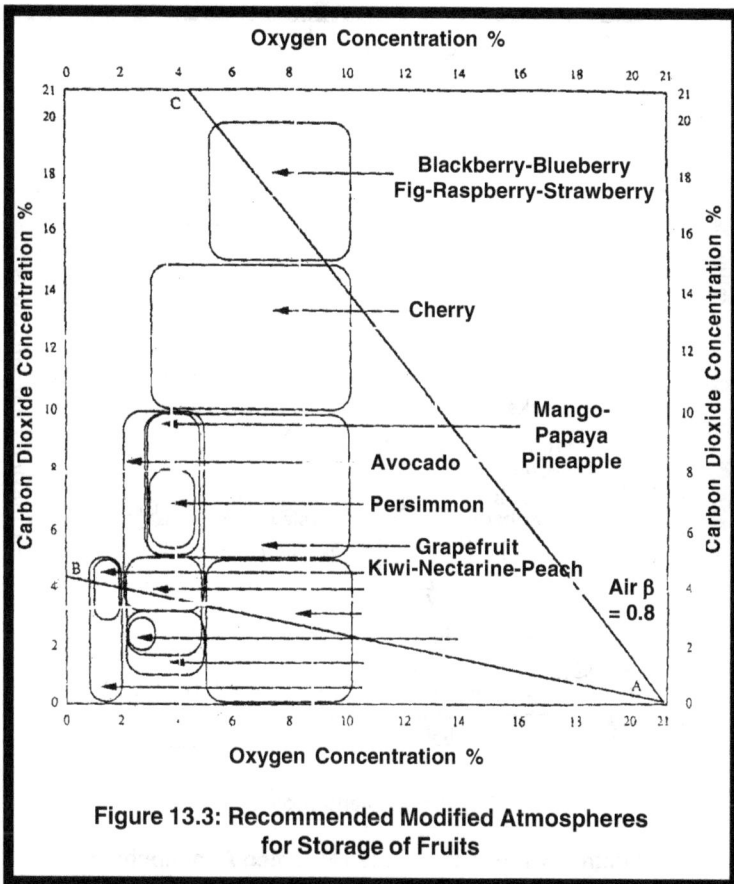

Figure 13.3: Recommended Modified Atmospheres for Storage of Fruits

maintain a desired atmosphere within a package. Details of such systems will now be discussed.

13.3 Package System for MAP

Polymeric films have been used for the packaging of horticultural products since the 1930s. Early work in the area stressed the primary role of packaging to reduce transpiration with many studies encouraging film perforation to avoid the development of injurious atmospheres in packages. Today interested centers on the development of non-perforated polymeric films which have the desired permeability characteristics so that modified atmospheres

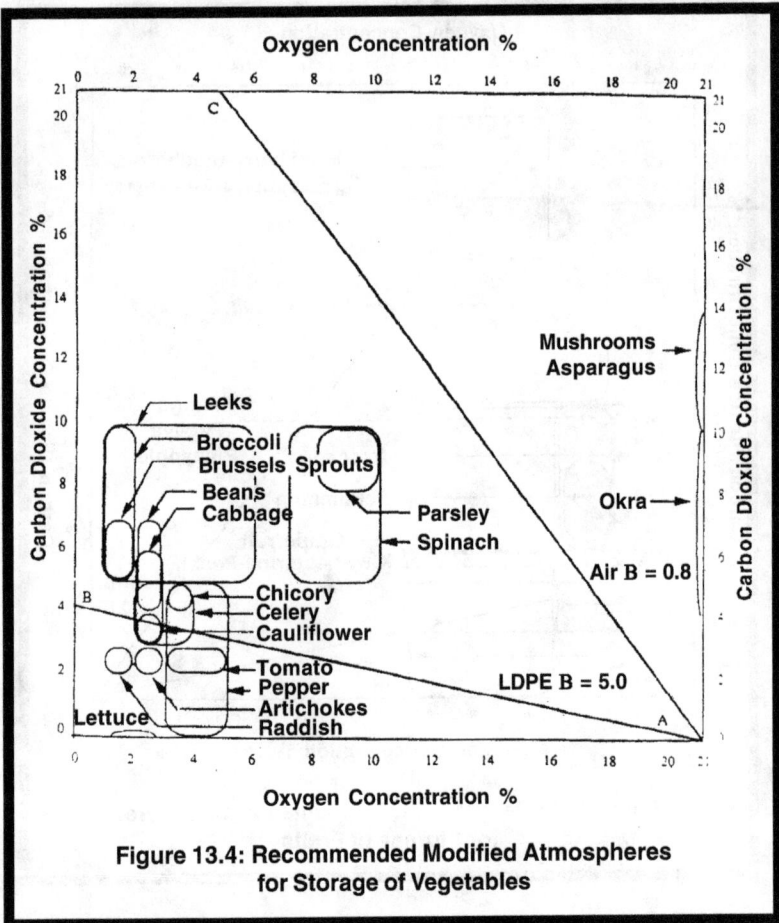

**Figure 13.4: Recommended Modified Atmospheres
for Storage of Vegetables**

appropriate for the particular commodity and its storage temperature
can be developed and maintained. In addition, the film must have
the ability to cope with fluctuating storage conditions (*e.g.,*
temperature, humidity and light) without injurious atmospheres
developing. It is important to note that in the presence of light, green
tissues undergo photosynthesis. In this situation, most of the CO_2
produced by respiration will be re-absorbed by photosynthesis, with
the result that the O_2 concentration in the atmosphere around the
produce will increase. This in turn will increase the rate of
respiration.

Although many polymeric films are available for packaging purposes, relatively few have been used to package fresh produce. This is not surprising since few have gas permeabilities that make them suitable to use for MAP. Those most likely to be suitable include LDPE, plasticized PVC, PS and EVA copolymer. The O_2 and CO_2 permeabilities of these films is required at actual conditions of use, but regrettably such data is scanty, making the development of suitable packages very much a "pack and pray" exercise in many cases. It is to be hoped that detailed permeability data (including the activation energies for permeation, diffusion and solubility, as well as parameters such as degree of crystallinity) will become widely available to assist package developers.

There are some fruits and vegetables for which LDPE film will never be suitable if a MAP is required. What is needed is the presentation of data for other polymeric films in a similar format to that for LDPE to assist in the selection of appropriate films and avoid wasteful trials with films which will never give the MAs desired. There is also a need for modification of the permeability properties of the common polymeric films to make them more suitable for MAP; research in this field is currently being undertaken involving the incorporation of inert, inorganic material into the film during extrusion.

A variety of configurations is available, ranging from a plastic bag which is heat sealed (sometimes after gas flushing), to a rigid, largely impermeable tray (either thermoformed from typically PET, PVC or PS, or made from molded paper pulp with a thermoformed plastic liner) to which is sealed a polymeric film having the desired permeability characteristics so that either an optimum MA is created inside the package at equilibrium, or a gas mixture flushed through the package immediately prior to sealing is retained.

During 1960s individual seal packaging system was developed. It is known as UNIPACK. Individual seal packaging (ISP) consists of packaging individual units of fruit in a polymeric film (originally 10 micron HDPE but now coextruded polyolefins) which is then shrunk by blowing hot air over the package. The fruit may be cooled after passing through the hot air tunnel by rapid ventilation. Equipment for seal packaging has developed rapidly, from hand wrapping, through manual, semi- and fully-automatic sealers, to automatic form-fill-seal machinery capable of operating at 200 fruit per minute. It is also possible to code or brand each individual fruit. The use of stretch rather than shrink film has also been attempted but there appear to be greater problems with sealing and mechanization of the process.

ISP appears to function primarily by maintenance of a water saturated atmosphere that extends shelf life by reduction of shrinkage and weight loss. It also reduces disease severity by preventing cross contamination of produce in boxes and permitting easy removal of infected individual fruit. The modification of respiratory gases within ISP produce is generally not sufficient to cause significant atmosphere modification.

A marked reduction in shrinkage and weight loss without a deleterious effect on flavor has been shown to result from ISP. It doubles and at times triples the shelf life of the fruit as measured by appearance, firmness, shrinkage, weight loss and other quality attributes. ISP also delays various parameters of physiological deterioration better than cooling to optimal temperature.

ISP has been shown to reduce chilling injury in various citrus cultivars, and thus, when combined with cooling which reduces decay, fruit deterioration is reduced. ISP also contributes to the healing of wounds caused during harvesting and markedly reduces the incidence of blemishes on various fruit. Changes related to senescence such as deterioration of cell membrane integrity and softening are also delayed by ISP. Undesirable effects of ISP include germination of seeds inside grapefruit stored for a prolonged time, sprouting of carrots, and rooting of citron fruit.

The use of films for ISP which contain fungicides that are gradually released has been suggested as advantageous because most of the fungicides is in the film and is not measured when toxic residues in fruit are examined. New films which contain substances capable of absorbing ethylene would also be excellent for ISP since reduction of the ethylene concentration, around the fruit would extend shelf life by delaying the various reactions leading to sensecence. For the degreening of citrus fruit, the application prior to sealing of ethylene releasing agents ensures that ethylene accumulates to a level suitable for accelerating the degreening process. There are advantages over the conventional degreening techniques where ethylene gas aggravates the development of various blemishes resulting at times in the loss of more than half the fruit.

Despite all the advantages listed above, the commercial adoption of ISP has been slow, possibly due to the lack of availability of high speed, reasonably priced packaging and equipment. However, the technique appears to be used extensively in China and Japan where ISP is performed by hand.

We can notice number of differences betweens CA storage and MAP of horticultural produce.

Despite the gas composition around the produce being almost identical in both cases, the storage life of MAP produce is frequently less than that of produce under CA storage. The limiting factor is the appearance of molds and bacteria in MAP produce as a result of the high (near saturation) in-package relative humidity (IPRH), a consequence of the low water vapor transmission rate of the films typically used for MAP. Such a problem is not observed in CA storage where the flow system presumably results in lower relative humidities.

Although fungicides have been used in an attempt to control mold development in produce packages they have been only partially effective and, in some cases, have caused problems of their own. The incorporation of fungicides directly into films has not been fully effective either. However, even if fungicides were effective, they do not control bacterial development which is also associated with high humidities in MAP.

The optimum RH for storage of fresh produce has not been well established, and the generally recommended levels of 85–95 per cent, represent a compromise between preventing excessive weight loss while providing some control of microbial spoilage.

There are two possible approaches to controlling IPRH: perforation of the package which precludes the possibility of achieving MA conditions within the package, and the use of in-package water-adsorbing compounds which can maintain a target RH. The feasibility of the latter approach to produce predetermined humidities within fresh produce packages has recently been described using predried crystals of compounds possessing a type III sorption isotherm according to the BET classification system.

The use of xylitol, potassium chloride, sodium chloride and sorbitol produced and held the IPRH at 78–79, 84–85, 73–76 and 72–74 per cent respectively, of a model package containing tomatoes and an LDPE film for gas exchange. No shriveling of the fruits was observed provided that their weight loss was less than 7 per cent. As well, because of the reduced RH and a_w inside the packages, microbial growth will not occur. Clearly, the use of potassium chloride would not be desirable since at that RH (0.84–0.85 a_w), most of the tomato fruit pathogens would be capable of growth.

In commercial packages, it is suggested that the salts be placed inside sachets of spunbonded olefin films which have small holes that readily permit the movement of water vapor but completely block the movement of liquid water. The critical feature of such a system is to have sufficient salt present such that when the desired quantity of water vapor has been absorbed, the salt remains saturated, for if the salt solution becomes unsaturated, the IPRH will rise. However, if the salt absorbs too much water vapor from inside the package, shriveling of the produce will result. It should be possible to design sachets having the quantity of salt which will achieve the optimum results.

13.4 Minimally Processed Fruits and Vegetables

Minimally processed fruits and vegetables (MPFV) are products that have the attributes of convenience and fresh-like quality, their forms varying widely depending on the nature of the unprocessed commodity and how it is normally consumed. The primary spoilage mechanisms are the metabolism of the tissue and microbial growth; both will cause deterioration of the tissue and must be controlled to maintain tissue viability.

Mechanical damage to cells during processing is a major limitation to shelf life of minimally processed fruits and vegetables. It has been shown that keeping shredded lettuce cold, using sharp slicing knives, minimizing cellular damage, eliminating cellular fluids, removing liquid containing active enzymes from the surface by centrifugation, and reducing the initial microbial population all contribute to longer shelf life. However, other workers have reported that washing shredded lettuce in deionized water and then spin drying before packaging did not improve the appearance over lettuce which had not been so treated.

Of key importance with MPFV is the control of enzymes, either endogenous from the produce itself or exogenous from the invading microorganisms to maintain the firm, crisp texture and bright, light color. The enzymes can be controlled either by inactivation or chemical or physical means. Methods that are currently used have been discussed and include temperature, sulfite solutions, modified atmospheres, ascorbic acid compounds, divalent ions very high pressures, additives such as L-cysteine (which inactivates polyphenoloxidase), vanillin (which inhibits polygalacturonase), mannose (which reduces ethylene production, respiration and softening in pears), and gases such as SO_2, CO and ethylene oxide. However, it is unlikely that the widespread use of the latter three gases would ever gain regulatory approval given the current trend towards additive-free food.

The use of a gas-exchange process for extending the shelf life of raw foods, and in particular, minimally processed apples and potatoes, was heralded in 1980 but appears not to have progressed to the commercial stage. The process consisted of the replacement of intra-tissue gases in particulate raw foods with one or more gases, in appropriate sequence, that stabilized or atleast significantly extended the like-fresh shelf life of the foods at ambient and/or chilled temperatures. Major savings in energy usage for the process of more than 75 per cent of that required for canning or freezing were also claimed.

Microbial deterioration of MPFV can be controlled by several methods including the use of chill temperatures, reduction of the total microbial population by the use of heat or irradiation, and the use of antagonistic organisms that control growth of undesirable microorganisms.

Very little has been published on the packaging of MPFV, but generally the same packaging processes as for fresh produce are used. However, allowance has to be made for differences in the respiration rate of produce which has been processed in some way, for example, the respiration rate of shredded lettuce is more than double that of whole lettuce quarters at 10°C, due to increased exposure of respiring tissue surface and physical damage causing a sound response. Thus, shredded lettuce will obviously need a considerably more permeable film than whole lettuce to result in aerobic equilibrium atmospheres.

The addition of an O_2 scavenger into an anaerobic package containing apricot and peach halves reduced textural loss from structural polymer breakdown, although a similar effect was not observed with pear halves. Scavenging the headspace O_2 from the package retarded textural breakdown to an even greater extent than the use of MA, although no details of the atmosphere were given. The optimum treatment for peeled and halved peaches and apricot halves consisted of a 2 per cent calcium chloride/1 per cent zinc chloride dip followed by anaerobic packaging and storage at 0–2°C; similar treatment did not retard degradation of pears.

The use of an ethylene absorbent (charcoal with palladium chloride) prevented the accumulation of ethylene and was effective in reducing the rate of softening in peeled and sliced kiwifruit and sectioned bananas, as well as chlorophyll loss in spinach leaves but not in broccoli.

Clearly, there is likely to be an increasing trend away from heavily processed towards minimally processed fruits and vegetables and considerable research will be required to develop the most effective packaging systems.

Packaging materials for frozen fruits and vegetables must protect the product from moisture loss, light and oxygen, of which the former is the most important. Freezer burn or sublimation of water vapor from the surface of frozen foods results in their becoming dehydrated with a concomitant loss in weight and their visual appearance deteriorates. All the common polymeric films have satisfactory water vapor transmission rates at freezer temperatures.

The earliest form of packaging material for frozen fruits and vegetables was waxed cartonboard, often with a moisture-proof regenerated cellulose film overwrap. These were then replaced with folding cartons having a hot melt coating of PVC/PVdC copolymer and the ability for the flaps to be heat sealed. Although still used to a small extent for low production volumes, the majority of frozen fruits and vegetables today are packaged in polymeric films based on blends of polyolefins, the major component of which is LDPE. Sufficient plasticizer is added to ensure that the films retain their flexibility, at low temperature. It is also common for the film to contain a white pigment to protect the contents from light which could oxidize the pigments. The film is usually supplied in roll form from which it is converted into a tube, then filled and sealed continuously in a form/fill/seal type of machine. Premade bags are used for low volume packaging operations.

The thermal processes used for canned fruits and vegetables differ markedly depending on the pH of the product: low acid products, *i.e.,* productions with a pH greater than 4.5 (which includes most vegetables) require a full 12D processes, typically 60–90 min at 121°C. In contrasts, those products with a pH less than 4.5 need only a mild heat treatment, typically 20 min in boiling water. Some products are acidified to lower the pH below 4.5 and thus, avoid the more severe heat treatment. For details about the processing procedures for individual fruits and vegetables, the reader is referred to a standard text in the area.

Canned fruits and vegetables are packaged either in tinplate cans or glass jars. The former must have the correct internal enamel applied to avoid corrosion of the tinplate. As with juices, it is important that all the air is removed from the product prior to packaging to minimize corrosion of the tinplate. For acid fruits containing anthocyanin red/blue pigments such as raspberries, the enamel coating must be particularly rigorous since the pigments act as depolarizers, accelerating the rate of corrosion. With some fruits only the ends of the can are enameled, and for pineapple a plain can is used; as the tin dissolves from the tinplate and reacts with certain constituents of the pineapple, a yellow color develops. White aluminum-pigmented epoxy resin enamels are used with fruits in some countries.

Many vegetables contain sulfur compounds which can break down during heat processing to release hydrogen sulfide. This can react with the tin and iron of the metal can to form black metallic sulfides which cause an unsightly 'staining of the can and also of the contents. While this process is encouraged and indeed is essential for the production of the desired flavor and color in canned asparagus, it is avoided with other vegetables by the use of special enamels which contain zinc oxide. This reacts with the hydrogen sulfide to produce barely detectable white zinc sulfide on the inner surface of the can. White aluminum-pigmented enamels based on epoxy resins are also used for cans containing vegetables.

Glass containers are still used for packaging some commercially processed fruits and vegetables, generally for products at the premium end of the market. This is largely because the production rates for glass containers are much lower than those possible for metal cans. Cylindrical, widemouth glass jars are commonly used with either a twist-off or pry-off cap made from mild steel or aluminum coated with a suitable lining material. Considerably greater operator skills are required to retort glass jars compared to metal cans, since failure to control the overpressure correctly can result in either shattered containers or the loss of pry-off caps.

Retortable pouches made from laminates of plastic film generally with an aluminum central layer can also be used for the packaging of fruits and vegetables which are preserved by the use of heat.

The packaging of dehydrated fruits and vegetables requires the use of a package which will prevent or at the very least minimize the ingress of moisture and, in certain instances, oxygen. For example, products which contain carotenoid pigments (*e.g.*, carrots and apricots) can undergo oxidative deterioration, and dehydrated potatoes are liable to develop stale rancid flavors unless oxygen is excluded. Vacuum or inert gas packaging may be used if the product is particularly sensitive to oxidation.

Many vegetables, for example green beans, peas and cabbage, are treated with sulfur dioxide prior to drying to retard nonenzymic browning (the principal cause of deterioration in dehydrated vegetables) and increase the retention of ascorbic acid. Sulfur dioxide also has a useful antimicrobial effect during the initial stages of drying and, by varying the form in which it is introduced (sodium sulfite or metabisulfite), it can be used to control pH, which in turn influences the color and subsequent handling and drying characteristics of the product. Concentrations of sulfur dioxide in the dried product normally range from between 200 and 500 ppm for potatoes to between 2000 and 2500 ppm for cabbage.

For the packaging of dehydrated fruits and vegetables, the material normally used consists of one or more polymeric films having the desired barrier properties. This implies that the material must be a very good barrier to water vapor and, depending on the particular product, a good barrier to O_2 and maybe SO_2 and certain volatiles.

Chapter 14
Packaging of Meat and Meat Products

14.1 Introduction

There is a continuous search for improved methods of transporting food products from producers to consumers. With the increasing urbanization of society, the problems associated with the keeping quality of fresh flesh foods have become more accentuated. Large livestock slaughter and processing facilities have developed in areas where livestock production is highly concentrated, and such areas are, not surprisingly, *well* away from the centers of population density. As well, there is a major world trade in fresh and preserved flesh foods. In all these situations, packaging has a key role to play in protecting the product from extrinsic environmental influences and giving the food the shelf life required in the particular market.

In the packaging of red meat, there are two factors of major importance: color and microbiology. Both of these will be considered in some detail before specific packaging materials and systems are discussed, since an understanding of red meat color and microbiology is an essential prerequisite to the development of successful packaging for red meat.

Since the water activity of chilled meat is very high, unprotected meat will lose weight by evaporation and its appearance will deteriorate. Further weight loss will occur when meat is cut, since the exposed surfaces exude liquid which detracts from the appearance of packaged meat. This can be overcome by including an absorbent pad in the base of the package. Although efficient chilling can reduce the quantity of exudate, a certain amount will always be present when meat cuts are held for retailing. This unattractive bloody exudate found in vacuum packaged beef is referred to as "purge", "weep" and "drip"; 1–2 per cent purge is considered acceptable, while 4 per cent is considered excessive. Values of 2–4 per cent drip can have substantial economic implications if not controlled.

The importance of color as a marketing attribute of red meat is well established, especially for self-service retailing. Consumers, used to seeing bright red meat prepared for sale, associate this color with good eating quality, although there is little correlation between the two. This association of the color of red meat (both in the chilled and frozen form) with freshness has been the dominant factor underlying retail meat marketing. The loss of this bright red color, known as loss of "bloom" in the industry, is affected by many factors, although the consumer will usually relate the color loss to bacterial growth. Because of the importance of meat color, various methods of transportation, distribution and packaging of primals and subprimals have evolved which optimize the maintenance of a desirable meat color.

The color of meat as perceived by the consumer is primarily determined by such factors as the concentration and chemical form of the meat pigment myoglobin, the morphology of the muscle structure and the ability of the muscle to absorb or scatter incident light.

Myoglobin is the principal pigment of fresh meat and the form that it takes is of prime importance in determining the color of the meat. The myoglobin molecule consists of a heme nucleus attached to a globulin type protein component. The heme group consists of a flat porphyrin ring with a central iron atom which has six bonding points or coordination links. Four of these are linked to nitrogen atoms; one is attached to the globin molecule; the remaining linkage is free to bind to other substances, usually water or oxygen. The color of myoglobin depends on at least three factors:

1. The oxidation state of the iron atom: it may be reduced or oxidized;
2. The nature of the group at the sixth bonding point of iron;
3. The state of the globin: it may be native as in raw meat, or denatured as in cooked meat.

The color of fresh meat depends chiefly on the relative amounts of the three pigment derivatives of myoglobin present at the surface: reduced myoglobin (Mb), oxymyoglobin (O_2Mb) and met myoglobin (MetMb). Reduced myoglobin (Mb) is purple in color and predominates in the absence of oxygen. Oxymyoglobin (O_2Mb) is bright red in color and results when Mb is oxygenated or exposed to oxygen; this is commonly known as "bloom." Metmyoglobin (MetMb) is brown in color and exists when the oxygen concentration is between 0.5 per cent and 1 per cent, or when meat is exposed to air for long periods of time. The brown metmyoglobin cannot bind oxygen even though it is oxidized by the same oxygen that converts myoglobin to the bright red oxymyoglobin.

The color reactions of myoglobin are all reversible and dynamic with respect to the three primary forms of myoglobin: Mb, O_2Mb and $MetMb^+$. It is rare when all the myoglobin color forms in meat are in the same form; generally two or more of the pigments will be present but the predominant pigment will be the most noticeable. $MetMb^+$ cannot take up oxygen but enzymes present in fresh meat are capable of reducing $MetMb^+$ to Mb which can then take up oxygen to form O_2Mb. As meat ages, the substrate for these enzymes is gradually used up and MetMb can no longer be reduced.

14.2 The Change of Color of Meat

For meat which has been exposed to air for several hours the penetration; depth may be 6–7 mm. The bright red color of O_2Mb will predominate and be apparent to the observer from the outside in to the point where the ratio of oxymyoglobin to myoglobin is about 1 : 1, *i.e.*, about 84 per cent of the total depth of O_2 penetration. Since different muscles have different inherent surviving respiratory

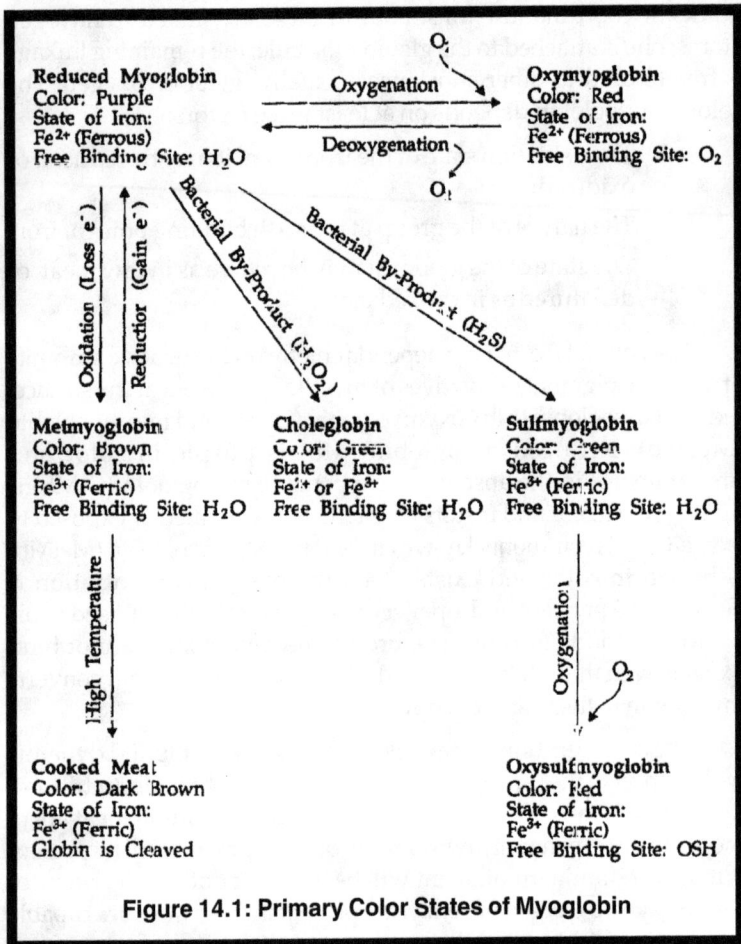

Reduced Myoglobin
Color: Purple
State of Iron:
Fe²⁺ (Ferrous)
Free Binding Site: H₂O

Oxygenation

Deoxygenation

Oxymyoglobin
Color: Red
State of Iron:
Fe²⁺ (Ferrous)
Free Binding Site: O₂

Oxidation (Loss e⁻)

Reduction (Gain e⁻)

Bacterial By-Product

Bacterial By-Product (H₂S)

Metmyoglobin
Color: Brown
State of Iron:
Fe³⁺ (Ferric)
Free Binding Site: H₂O

Choleglobin
Color: Green
State of Iron:
Fe²⁺ or Fe³⁺
Free Binding Site: H₂O

Sulfmyoglobin
Color: Green
State of Iron:
Fe³⁺ (Ferric)
Free Binding Site: H₂O

High Temperature

Oxygenation

Cooked Meat
Color: Dark Brown
State of Iron:
Fe³⁺ (Ferric)
Globin is Cleaved

Oxysulfmyoglobin
Color: Red
State of Iron:
Fe³⁺ (Ferric)
Free Binding Site: OSH

Figure 14.1: Primary Color States of Myoglobin

activities, X for different muscles will vary by a factor of 2–3 under a given set of conditions.

Due to a lack of oxygen, the color of the interior of fresh red meat is purple because of the presence of reduced myoglobin. The outer most layer exposed to air is red due to the presence of oxymyoglobin. A thin brown layer of metmyoglobin can be observed between the red and purple regions due to a low concentration of oxygen. This layer thickens within for 1 or 2 days and becomes apparent, first by

Zero Oxygen
Pressure

Oxidizing Agents
Low O_2 Pressure

Oxymyoglobin ⟶ Reduced Myoglobin ⟶ Metmyglobin

$Fe^{2+} \cdot O_2$ ⟶ Fe^{2+} ⟶ Fe^{3+}

High O_2 Pressure   Reducing Agents and Enzymes

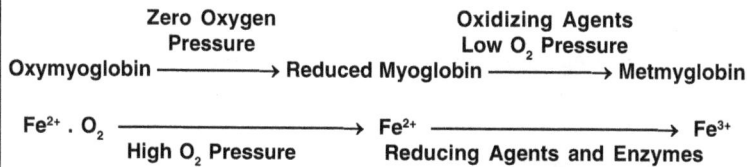

Figure 14.2: Schematic Diagram of Interrelationships Between Three Major Meat Pigments

darkening the translucent surface tissue and later by breaking through to the surface. At lower oxygen partial pressures, the MetMb layer will be nearer the surface until at the critical partial pressure it is at the surface. The rate of these changes is strongly influenced by the temperature.

Oxygenation of purple reduced myoglobin to red oxymyoglobin is rapid, the surface of beef in air appearing red within half an hour at 5°C. In contrast, oxidation to metmyoglobin is slow, appearing first as a thin brown layer at the limit of oxygen penetration. The red oxymyoglobin is stable as long as the heme remains oxygenated, although the oxygen is continually associating and dissociating from the heme.

The dissociation of oxygen from the heme (known as deoxygenation) is caused by conditions such as low pH (less than pH 5.4), high temperature, ultraviolet light, salts and especially low oxygen tensions which cause denaturation of the globin moiety of oxymyoglobin. The deoxygenation of red oxymyoglobin results in myoglobin (which is very unstable) which then becomes oxidized to brown MetMb. Surface dessication increases the salt concentration and promotes the formation of MetMb.

The conditions which cause deoxygenation of oxymyoglobin to myoglobin are also responsible for the conversion of unstable reduced myoglobin to brown metmyoglobin. Since the globin moiety is denatured, the oxidation (often called autoxidation) proceeds spontaneously and involves the loss of an electron from the iron of the heme. The actual oxidation is caused by the activity of oxygen-utilizing enzymes indigenous to muscle tissue. Through a process of respiration the O_2 is converted to CO_2. When the O_2 has been substantially reduced, the MetMb pigments are reduced to the purple Mb.

Low storage temperatures suppress the activity of these enzymes, but with increasing storage temperatures, their activity increases thus, lowering the oxygen tension at the meat surface. Lamb muscles have the highest concentration of these enzymes compared to beef and pork and therefore, a very high oxygen demand; this explains the comparatively short retail shelf life of lamb. As well as encouraging enzyme activity, high storage temperatures will also decrease the depth of oxygen penetration into the meat tissues, moving the brown met myoglobin layer closer to the surface.

The maximum rate of metmyoglobin formation has been reported at oxygen partial pressures of 7.5 ± 3 mm Hg at 7°C and 6 ± 3 mm Hg at 0°C for beef. Earlier results which are still frequently quoted are less relevant since they were obtained using pure pigments and at temperatures well above those likely to be encountered in practice. For example, the rate of autoxidation for pure myoglobin solutions was reported as maximal at about 1 mm Hg of oxygen at 30°C and pH 5.7, while the maximal rate for ox blood hemoglobin at 25°C and pH 5.7 was reported to be 20 mm Hg of oxygen. Clearly, any value is very dependent on temperature, although independent of pH. If meat is held under high O_2 partial pressures (*i.e.*, above 30 mm Hg), autoxidation of Mb is minimized. These results have led to the development of systems allowing transport of meat at high O_2 partial pressures.

Metmyoglobin reduction to myoglobin in *post rigor* meat (*i.e.*, a change in the valence state of the iron molecule from ferric to ferrous) is primarily enzymic in nature, with an electron being donated by reducing enzyme systems. This process is often called "MRA" or "metmyoglobin reducing activity." Some muscles will remain bright red for longer periods of time than other muscles because any metmyoglobin formed is reduced back to reduced myoglobin and immediately oxygenated back to red oxymyoglobin; such muscles are said to have a higher MRA. Loss of reducing activity in meat *post mortem* is due to a combination of factors including fall in tissue pH, depletion of required substrates and co-factors, and ultimately complete loss of structural integrity and functional properties of the mitochondria.

It has been reported that bacteria cause discoloration of meat in their logarithmic growth phase, apparently due to the high oxygen demand of aerobic bacteria reducing the oxygen tension at the meat surface and causing the formation of metmyoglobin.

Some bacteria also produce by-products which oxidize the iron molecule. The most common are hydrogen sulfide (H_2S) and hydrogen peroxide (H_2O_2) which react with unstable myoglobin to produce sulfmyoglobin and choleglobin, respectively. H_2S production causes green discoloration on vacuum packaged meat, it generally being found only on meat having a high pH (pH > 6). On opening a vacuum packaged cut that has green discolorations on it, the green sulfmyoglobin becomes oxygenated to oxysulfmyoglobin which is red in appearance. However, the color change does not cause any diminution in the H_2S off-odor which resembles rotten eggs. H_2O_2 can either cause greening due to oxidation of heme pigments to form choleglobin or degradation beyond porphyrins to bile pigments.

The color intensity of meat is determined by *ante mortem* factors such as species, stress, sex and age of the animal, the *post mortem* pH rate of decline and the ultimate pH of the meat. During *post mortem* glycolysis the pH of normal tissue falls from an *in vivo* value of 7.2 to an ultimate value of 5.4–5.6. As well as influencing the color, the rate of fall of pH and the ultimate pH value also influence the water holding capacity and the texture of the meat. Differences in color intensity between species are primarily caused by differing concentrations of myoglobin. Thus, beef which has the highest concentration is the darkest of the meat species, with lamb being intermediate in color and myoglobin concentration. Pork has the lowest concentration of myoglobin and as such is the lightest in color. Male animals usually produce darker meat than females due to a greater concentration of myoglobin in meat derived from male animals.

The color intensity of meat is also influenced by pH and morphology of the muscle structure. The pH of a muscle is largely influenced by the conditions which exist immediately prior to or just after slaughter. Short term, violent excitement immediately before slaughter, or the slow cooling of carcasses after slaughter, can result in meat having a low ultimate pH. This is caused by a buildup of lactic acid which is formed as a metabolic end product in the anaerobic breakdown of glucose and glycogen. This condition is known as PSE since it produces pale, soft and exudative meat as a result of an abnormally rapid fall in pH immediately after slaughter when the combined effects of low pH and high temperature in the muscle lead to denaturation of sarcoplasmic and myofibrillar proteins. Because muscle glycolysis is relatively more rapid in hogs than bovines, PSE is rarely observed in beef animals. However, there are appreciable differences in beef with respect to both color and water-holding capacity due to differences in the temperature/pH profiles postmortem. It would appear that PSE meat spoils in the same manner as nonnal meat.

PSE meat causes problems in packaging: the color is abnormally pale due to denatured protein, the physical structure increases light scattering making the meat more opaque and autoxidation increases causing color fading. In addition, because of its low water-holding capacity resulting from protein denaturation, PSE meat produces excessive drip.

Biochemical conditions directly opposite to those producing PSE meat give rise to another type of abnormal meat from all meat animals but more especially beef cattle and hogs. This condition has been described as even more troublesome than PSE from a packaging point of view. The problem meat, known as dark-cutting or dark, firm dry (DFD), is translucent and sticky to touch, and unacceptable for retail packaging because of its dark purple color. This meat has a high pH due to a low residual glycogen level remaining in the muscle at slaughter, following excessive stress (such as an extended transit haul) or exercise prior to slaughter. Such treatment causes depletion of the muscle glycogen before death. The low residual glycogen means that insufficient lactic acid is formed by glycolysis during the rigor process to lower the pH.

The muscle fibers of DRD meat are swollen and tightly packed together forming a barrier to the diffusion of oxygen and the absorption of light. As well, the high pH of the meat accelerates respiratory activity of the meat tissue, resulting in a very thin layer of red oxymyoglobin with the underlying purple reduced myoglobin more visually apparent. Moreover, the high pH encourages the growth of putrefactive microorganisms, thereby significantly reducing the keeping quality of this meat. For example, vacuum packaged beef of normal pH can be stored at chiller temperatures for periods in excess of 10 weeks, whereas DRD beef held under the same conditions will generally spoil within about 6 weeks. However, both normal and high pH cuts of beef packaged under CO_2 were unspoiled at 15 weeks.

The effect of carbon dioxide on meat color is questionable. It has been reported that meat stored in high concentrations of CO_2 often develops a grayish tinge supposedly because of the lowering of the pH and subsequent precipitation of some sarcoplasmic proteins. It was recommended that CO_2 not be used in concentrations exceeding 20 per cent, and the claim was made that no MetMb formation occurred in atmospheres containing high concentrations of CO_2 providing that the O_2 partial pressure exceeded 5 per cent. A later report indicated that 50–80 per cent CO_2 is often found in residual air spaces in vacuum packages with no associated detrimental effect on meat color.

More recently, the realization that discoloration in high CO_2 atmospheres must be due to oxygen leading to the use of 100 per cent CO_2 atmospheres on the grounds that meat color would not be adversely affected if oxygen was rigorously excluded from the packages. In the case of raw red meats, the color was actually improved by enzymic reduction of pigment that oxidized before the meat was packaged. However, the stability of raw meat color on exposure, to air decreased as storage life increased, the color stability at display reaching a minimum value after 12 weeks storage. After that time, acceptable meat color at display was maintained for about 40 per cent of the time that fresh meat retained acceptable color. This was claimed to be an inevitable consequence of prolonged chill temperature storage and not an effect of the 100 per cent CO_2 atmosphere.

Carbon monoxide (CO) has the potential to retard MetMb formation and fat oxidation. CO combines with myoglobin to form the bright red pigment carboxymyoglobin (COMb) which is spectrally very similar to oxymyoglobin (O_2Mb). Since COMb is much more stable toward oxidation than O_2Mb by virtue of the stronger association of CO to the iron-porphyrin site on the myoglobin molecule, the addition of CO at low levels can act to negate the detrimental color changes associated with the high levels of CO_2 required to maintain wholesomeness over prolonged transit times.

Significant retardation of rancidity in frozen pork sausage stored under CO-containing atmospheres has been reported, and it has been shown that the presence of CO increases the proportion of Mb present in a reduced state.

Despite these advantages, CO has not been approved by regulatory agencies for commercial use with meats. This is largely because of safety considerations, although it has been shown that less than 0.09 ppm of CO residue was left on cooked ground beef which had been exposed to CO storage.

Lighting is an important factor when presenting meat and meat products for retail sale. A high percentage of red in the light gives a particularly intense, red impression, and products that are already beginning to turn a grayish-red in daylight appear saturated with red when exposed to reddish lighting.

Incident light is a contributory factor in the discoloration of both fresh and frozen meat as, for example, in a retail supermarket display case. The extent of the effect depends on such factors as the wavelength and intensity of the light, temperature, oxygen, partial pressure, meat pH and storage time. Provided that UV wavelengths are avoided, the effect of light is small under usual refrigerated conditions.

The color of frozen meat initially depends on the rate of freezing and the resultant size of ice crystals in the surface layer. Slow freezing produces large ice crystals with poor light scattering properties, giving the meat a dark, translucent appearance. Fast freezing, on the other hand, results in the formation of small ice crystals which scatter light and make the meat surface pale and opaque.

In frozen red meat, the principal deterioration during storage is photo-oxidation of the pigment. Whereas under direct illumination, chilled meat oxidation begins in the subsurface layer and progresses towards the surface, frozen meat oxidizes from the surface inwards. Loss of redness is detrimental to the marketing of frozen meat as well as chilled meat.

Considerable improvements in the color shelf life (in the order of 5–10 times) of prepackaged frozen beef have been reported when a light intensity of 100 lux was used compared to the normal retail cabinet intensity of 1000–2000 lux. Display lighting affected a graying and loss of saturation of the color of frozen lamb chops.

14.3 Packaging of Fresh Meat

In the USA prior to 1967, whole beef carcass shipments were traditionally made to retail outlets where carcasses, were fabricated into retail items by butchers in individual stores. The "boxed beef" concept was introduced in 1967, dramatically changing beef processing, distribution and retail fabrication, without affecting retail presentation to the consumer. The basis for the boxed beef concept was vacuum packaging in low oxygen barrier materials which provided a good barrier to oxygen.

It is now a common practice to butcher beef carcasses at the slaughter house into 2–9 kg primal joints which are then vacuum packaged into plastic bags with low permeabilities to gases. In these circumstances, the atmosphere around the meat becomes depleted in O_2 (often < 1 per cent v/v) and enriched in CO_2 (> 20 per cent v/v), resulting in microbial changes quite different from those observed during aerobic storage. Vacuum packaged boxed beef is then distributed to retail outlets where these primals and subprimals are fabricated into consumer units, overwrapped in oxygen permeable film on PS foam or PVC trays and displayed for sale.

Boxed beef developed rapidly and changed beef distribution dramatically in the USA. In addition to beef, vacuum packaged boxed pork and lamb followed in the 1970's. In 1986 approximately 84 per cent of all beef processed in the USA was vacuum packaged.

Vacuum packaging achieves its preservative effect by maintaining the product in an oxygen-deficient environment. In anoxic conditions, potent spoilage bacteria are severely or totally inhibited on low-pH (< 5.8) meat. However, their growth on high-pH muscle tissue, or extensive fat cover of inevitably neutral pH, will spoil relatively rapidly in a vacuum pack. Vacuum packaging can therefore, extend the shelf life of primal cuts composed largely of low (normal) pH muscle tissue such as beef and venison by about fivefold over that achieved in air. For other meats and small cuts, only a twofold extension of shelf life can be safely anticipated.

The introduction of vacuum packaging for the distribution and storage of chilled beef has been hailed as the greatest innovation in meat handling during the last 25 years. Of the 11 million tonnes of beef slaughtered in the USA in 1988, approximately 87 per cent moved into distribution in the form of boxed beef for industrial, restaurant and institutional use. Beef carcasses are broken down into primal and subprimal cuts, separated into boneless and bone-in cuts, and then vacuum packaged. Since only about two-thirds of a beef carcass is usable meat, there are clear advantages in the form of reduced refrigerated space for transportation and storage and less packaging material when boneless beef rather than bone-in or carcasses are stored and distributed. Another advantage is that the tenderness of the beef can be improved by aging without the evaporative weight loss incurred when carcasses are hung in the conventional manner.

Vacuum packaging involves enclosing large boneless joints (typically 3 to 15 kg in weight) in flexible plastic containers (usually bags) to prevent moisture; loss and exclude oxygen from the meat's surface. Packing under a vacuum reduces the volume of air sealed in with the meat. The plastic materials used for vacuum packaging must have low moisture and gas permeabilities and be strong enough to hold heavy beef joints.

In packaging fresh meat primals, many cuts contain bones which are often sharp and abrasive and readily puncture the flexible plastic materials used in vacuum packaging. To overcome this bone puncture a BONEGUARD material (approved by the FDA and USDA) consisting of a wax-impregnated and coated cotton scrim is employed.

Reported oxygen permeabilities of packaging films are usually measured at ambient temperatures and moderate humidities (typically 23°C and 50 per cent RH); but both temperature and humidity can affect the rates at which gases are transmitted through films. Data on the oxygen permeabilities of packaging films at chill temperatures are sparse, and those that do exist often do not include a complete specification of the film under test, or the test conditions. This is further complicated by the variety of test methods and units used. The oxygen permeabilities at sub-zero temperatures of two plastic films used for the vacuum packaging of meat have been reported. One film was a polyamide-polyethylene laminate, while the other was an EVA copolymer-PVC/PVdC copolymer laminate. Their oxygen permeabilities at -1°C were reported as 2.0 and 0.6 mL m^{-2} 24 h^{-1} atm^{-1} respectively, approximately one fiftieth of the values obtained at 23°C and 90 per cent RH.

As has been pointed out, once a piece of meat has been vacuum packaged in an oxygen barrier material having a gas permeability (at 23°C, and 75 per cent RH) below 50 mL m^{-2} day^{-1} atm^{-1} and adequately sealed to prevent air reentry, the shelf life of the meat is very much the same regardless of the packaging material. Therefore, the significant differences between the packaging materials and/or systems are not in the structure so much as in the physical properties, the production speeds of the system and the abuse resistance of the package itself.

When meat is vacuum packaged the contaminating flora are exposed to an atmosphere containing 20–25 per cent CO_2 and less than 10 per cent O_2. Both the high CO_2 and low O_2 tension depress the growth of Pseudomonads and facultative anaerobes predominate: lactic acid bacteria, *B. thermosphacta* and *Enterobacteriaceae*. The antimicrobial activities of the Lactobacilli, coupled with, the low

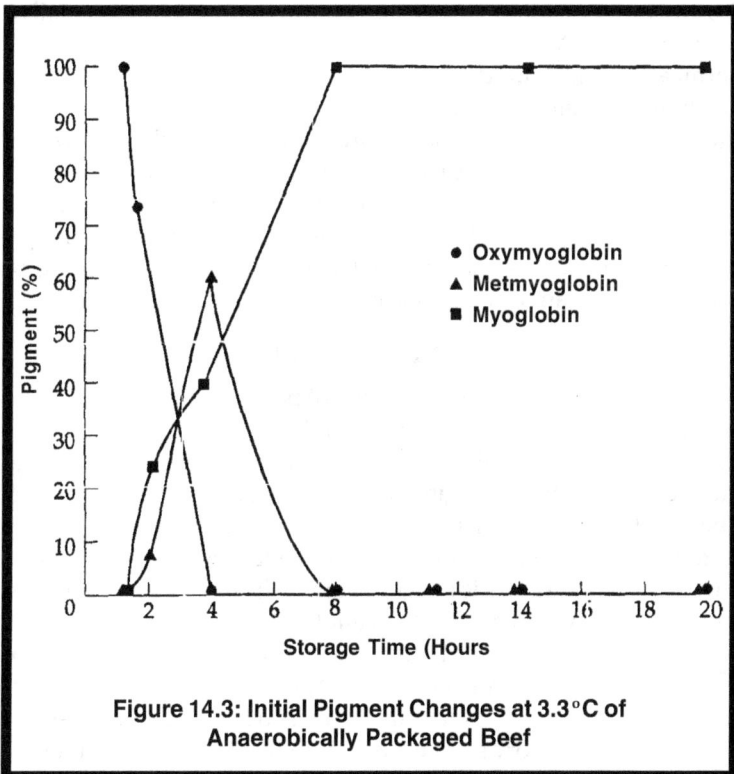

Figure 14.3: Initial Pigment Changes at 3.3°C of Anaerobically Packaged Beef

storage temperature, combine in a synergistic manner to extend the shelf life of vauum packaged meat.

The oxygen in the small volume of residual air inside a vacuum package will be quickly consumed by meat respiration so that within 2 days the oxygen partial pressure at the surface of the meat will have dropped below 10 mm Hg. At these very low partial pressures the penetration limit of oxygen is very near the surface and the thin brown layer of MetMb which develops cannot conceal the underlying Mb; therefore the visible color of vacuum packaged beef is purple.

The efficacy of vacuum packaging depends on there being close contact between a film of low gas permeability and all product surfaces. If there are vacuities within the pack, these will develop an oxygen-containing atmosphere as gases permeate into the package during storage. Bacterial growth will accelerate and product color will deteriorate because of oxidation of Mb at meat surfaces exposed to such an atmosphere. Also, when the meat surface is large relative to the meat mass, an extensive film surface is presented for oxygen permeation and color deterioration can also occur. Consequently, vacuum packaging is relatively ineffective for preserving products such as carcasses whose shapes preclude close application of the packaging film to all surfaces, and small cuts of any shape. Furthermore, because the anoxic conditions within the package result in the formation of purple Mb, vacuum packaging for retail display packages of red meat has not met with wide success because of consumers' negative perception of the color purple.

14.4 The Shelf Life of Packaged Meats

In any discussion of the shelf life of packaged meats, it must be borne in mind that comparisons of published data are difficult and the making of generalizations foolhardy since there are a large number of variables which interact to determine the actual shelf life. The most important of these variables is temperature, and the statement that samples of vacuum packaged meat were held at a particular temperature is not especially helpful unless the range of temperatures encountered by the samples during storage is specified. Generally it can be said that laboratory scale trials control temperatures over a much smaller range compared to commercial scale trials, and this is frequently the reason for the longer shelf life obtained in laboratory scale trials over that obtained in commercial production.

Other important variables include the microbiological status of the meat at the time of packing and the method used to determine the end of shelf life of the meat. These methods range from objective assessments of microbial counts to subjective assessments by taste panels consisting of trained or untrained members. Clearly, attempting to draw generally applicable conclusions from numerous published reports where the magnitude of these variables differs (or is not even specified) would only be misleading.

Vacuum packaged beef of normal pH can be stored at chiller temperatures for periods in excess of 10 weeks. However, high pH, dark, firm, dry (DRD) beef held under the same conditions will generally spoil within about 6 weeks. The degree of vacuum has been reported as having no effect on sensory panel ratings of vacuum packaged beef cuts. There are also advantages in aging beef in vacuum packages as opposed to carcass aging, including less loss due to water evaporation, less necessity for trimming of exposed surfaces and more efficient use of refrigerated space.

Although for many years vacuum packaging was only applied to chill beef, it has been applied in recent years to lamb and pork. Because of their relatively small size, pork and lamb carcasses are only partially boned before packaging, and the presence of bone can lead to puncturing of the package unless precautions are taken, for example, the use of BONEGUARD material (a woven cotton scrim impregnated with wax and having dead fold characteristics) or strong EVA copolymer film in the package structure.

In contrast to beef cuts, much of the surface of lamb cuts is adipose, rather than muscle tissue. Adipose tissue has pH values close to neutrality and has no significant respiratory activity. Packaged lamb can therefore, present a heterogeneous environment for microbial growth. This different microbial environment of high pH and, possibly, relatively high oxygen concentration, probably accounts for the shelf life of only 6 to 8 weeks reported for vacuum packaged lamb whereas 11 to 12 weeks is routinely attainable with beef. The longer shelf life of vacuum packaged beef is due to its flora being dominated by Lactobacilli which overgrow other spoilage organisms in the low temperature, low pH, low O_2 and high CO_2 environment of the package.

Laboratory and commercial trials in New Zealand seeking to extend the chill temperature shelf life of lamb cuts showed that after

12 weeks' storage at –0.5°C, lamb loins vacuum packaged in a PET-aluminum foil-LDPE laminate having an immeasurably low gas permeability developed floras of Lactobacilli that had not caused spoilage; meat color was also much improved by the exclusion of oxygen. Commercial trials where the packaged cuts were held for 6 weeks at 0°C and then at –0.5°C were very acceptable at 9 weeks but most were spoiled by off-flavors at 12 weeks, the next sampling time. Since then a high CO_2 MAP has been developed in New Zealand which gives a chill temperature shelf life for lamb of at least 16 weeks; there is now less interest in extending the chill temperature shelf life of vacuum packaged lamb.

Various conflicting reports have appeared in the literature concerning the chill temperature shelf life of pork joints. For example, figures of little more than 2 weeks at 1°C and 3–4 weeks at 2°C were reported in an extensive review. However, it appears that limited use is being made of vacuum packaging for wholesale cuts of pork, where the attraction is likely to be convenience in butchery rather than savings in weight loss.

Recently a comparison of three packaging treatments [MAP (25 per cent CO_2; 75 per cent O_2); VSP and the mother bag system (100 per cent CO_2)] on the shelf life of fresh pork stored at 0°C was reported. All three treatments were equally efficient for the first 4 days of retail display, but the mother bag system gave the most promising shelf life results (21 days). The MAP shelf life was 14 days and the VSP 7 days.

The chill temperature shelf life of vacuum packaged lamb, mutton and pork has been extended using a process developed in Australia which involves application of foodgrade acetic acid at a concentration of 1.5 per cent and a temperature of 53–55°C (higher temperatures may cause discoloration). This procedure results in an immediate reduction of 90–99.5 per cent in the bacteria on the exposed surfaces and it has a residual action. A 3 per cent solution (higher concentrations may cause adverse color changes) at 23–25°C has a similar effect. An increase in storage life of about 50 per cent was achieved using this procedure prior to vacuum packaging. Because the 1.5 per cent acetic acid solution at 55°C may cause irritation to personnel and possibly corrosion to metal surfaces, the more expensive 3.0 per cent solution used at the lower temperature is preferred. Both processes have the approval of the FDA and the USDA, but not the European Community.

14.5 The Poultry Meat and Eggs

Although most poultry meat is sold in the form of whole, oven-ready carcasses, there is now an increasing demand for cut-up portions and a variety of other further processed products, both raw and cooked.

Raw poultry meat is a perishable commodity of relatively high pH which readily supports the growth of microorganisms when stored under chill or ambient conditions. The shelf life of such meat depends on the combined effects of certain intrinsic and extrinsic factors, including the numbers and types of psychrotrophic spoilage organisms present initially, the storage temperature, muscle pH and type (red or white), as well as the kind of packaging material used and the gaseous environment of the product.

The main pathogenic organisms associated with poultry and poultry products are *Salmonella* spp.; *Staphylococcus aureus,* and *Clostridium perfringens.* Most studies on the extension of shelf life using CO_2 in modified atmospheres have concentrated on the suppression of spoilage organisms rather than the survival and growth of pathogens.

The vacuum packaging of poultry carcasses, cuts and other manufactured products can extend shelf life provided that the product is held under chill conditions. During storage at 1°C in either O_2 permeable film or vacuum packs, extensions in shelf life from 16 to 25 days in the case of breast fillets (pH 5.9–6.0) and from 14 to 20 days for drumsticks (pH 6.1–6.3) were observed for the vacuum packaged products. Somewhat surprisingly; deleterious flavor changes tended to precede the development of off-odors in vacuum packs of both types of muscle.

Although vacuum packaging may be used to encourage the development of an atmosphere around the product which delays microbial spoilage, other oysters in use involve an initial addition of about 20 per cent CO_2, to either individually packaged items or bulk packs of varying size in an O_2 impermeable plastic film. The MAP of poultry and related products has been reviewed.

The most comprehensive study of the use of CO_2-enriched atmospheres for extending the shelf life of poultry meat (chicken portions) was carried out in the USA in 1951. The maximum usable

CO_2 concentration was 25 per cent since above this the meat became discolored; even at 15 per cent a loss of bloom was sometimes noted.

Transferring CO_2-stored poultry meat to normal atmospheric conditions of cold storage gave an extended shelf life which was intermediate between the keeping time in air alone and that in the CO_2 atmosphere. Whether this effect is due to a lag phase in microbial growth induced by a change of gaseous, environment or merely to the slow diffusion of CO_2 from the tissues and hence a "residue" effect has not been established.

As was mentioned earlier in this chapter, CO_2 levels of 20 per cent or more extend; the shelf life by inhibiting the growth of many organisms at temperatures of $-1°C$ to $7°C$, the inhibitory effect being critically influenced by temperature. This was demonstrated graphically by a study into the effect of 80 per cent CO_2 (balance air) compared to air on the survival and growth of microorganisms most often associated with spoilage and foodborne disease in poultry carcasses at 2, 7 and $13°C$. The CO_2 atmosphere substantially retarded the growth of the total bacterial load in uninoculated ground chicken meat and parts at all temperatures when compared to air; however, temperature had a larger overall effect than atmosphere. The study concluded that MAP of refrigerated poultry in elevated CO_2 atmospheres does not increase the microbial hazards when compared to air at the same temperature.

Another study found that 100 per cent CO_2 markedly reduced the growth rate of microorganisms whereas the growth rate in 20 per cent CO_2 was only slightly less than that in vacuum packs. No adverse effects of high CO_2 concentrations on meat quality were reported.

Although there is evidence that CO_2 packaging can be more effective in extending the shelf life than vacuum packaging, most studies in which an 'appreciable shelf life extension has been demonstrated have involved CO_2 concentrations above the 25 per cent limit suggested by the early American work.

The optimum modified atmosphere to limit microbial growth and insure organoleptic quality to 27 days of storage at $1.1°C$ of whole chickens in a polyamide/ionomer pouch with a minimum replacement of 95 per cent of the air required a usage level of 722 mL of CO_2 per kg of whole chickens. Although atmospheres enriched with CO_2 have been advocated for extending the shelf life of processed poultry products, the precise composition of the atmospheres necessary to achieve the desired result have yet to be clarified in the published literature.

While the number and types of microorganisms found on stored poultry is an important factor when determining the shelf life, the real determinant is the sensory quality of the raw and cooked product. Unfortunately most published studies have not included sensory tests. One study which did evaluate the quality of raw and cooked poultry that had been stored under MA and refrigeration for up to 5 weeks. Their data indicated that MAP (80 per cent CO_2) poultry would be quite acceptable to consumers for up to 4–6 weeks depending on the temperature of storage. It was noted that commercial poultry processors may not get as long a shelf life because of difficulties in controlling the packaging process and temperature under production conditions.

In a study of broiler carcasses packaged under vacuum in film of raw O_2 permeability, or under CO_2 in gas-impermeable packages, shelf life was a function of storage temperature, packaging and O_2 availability. Putrid spoilage in gas-impermeable packages after 7 weeks storage at 3°C or 14 weeks storage at –1.5°C was attributed to Enterobacteria. In vacuum packages with oxygen transmission rates of 30–40 mL m^{-2} day^{-1}, putrid odors were detected after 2 weeks storage at 3°C and 3 week storage at –1.5°C.

Concerning the safety of MAP chicken, the possible problem organisms would be *C. jejuni* which may be able to survive better in a MAP product, and *L. monocytogenes* and *A. hydrophila* which may, because of the extended storage lives of the MAP products, have additional time to grow to potentially high numbers. Although *C. perfringens* may be able to survive better in some MAs as compared to air, it would not be able to grow at the chill temperatures commonly used for MAP products. Thus, it is unlikely to be a health hazard in a MAP product unless the product is temperature abused, because high numbers of the organism must be ingested to cause illness.

Handling poultry products during a postcook packaging operation can contribute more to bacterial spoilage than the packaging operation itself. It has been reported that turkey product packaging had only a minor effect on numbers of bacteria, the significant increase in bacterial count being due to the handling operation, probably from equipment surfaces.

Infertile eggs from hens classified as *Callus domesticus* are used almost exclusively for human consumption, the average weight of shell eggs being about 58 g. Shell eggs consist of 8–11 per cent shell, 56–61 per cent albumen, and 27–32 per cent yolk. The pores in the shell permit gaseous diffusion as well as moisture loss and the entry route for microorganisms which might infect the egg. Egg products are commonly marketed in the following forms: (a) refrigerated, (b) frozen, and (c) spray dried.

The factors associated with the loss of shell egg quality are time, temperature and humidity. During the storage of shell eggs, the pH of albumen increases at a temperature-dependent rate from about 7.6 to a maximum value of about 9.7, the rise resulting from a loss of CO_2 through the pores in the shell. The rise in pH of albumen causes a breakdown in the gel structure of the thick white.

Several methods of altering the environmental conditions around eggs have been used to prolong their shelf life, the major one being refrigeration. Coating of the shell with mineral oil has also come into common usage to increase the shelf life by reducing the rate of CO_2 and moisture loss. Spray oiling has come into common usage for the shells of eggs, light mineral oils of food quality normally being used. It is essential that the treatment be applied to the eggshell within a few hours of production since the rate of CO_2 loss is very high during the first few hours, and weight loss during the first few days can be significant.

Clearly the rate of CO_2 loss at any particular temperature will be a function of the partial pressure of CO_2 in the external environment. Thus, it is not surprising that attempts to extend the shelf life of shell eggs have involved storing the eggs in an atmosphere containing CO_2 or alternatively packaging eggs in a CO_2-impermeable package. A comparative study evaluating the quality and shelf life of fresh shell eggs stored at room temperature with four different treatments (not packaged; packaged in air; packaged with 15 per cent CO_2 and oil coated) concluded that controlled atmosphere was the most efficient method of preserving egg quality at room temperature for a period of 7 weeks.

Spoilage of flesh foods such as fish and shellfish results from changes brought about by chemical reactions such as oxidation, reactions due to the fish's own enzymes, and the metabolic activities of microorganisms. The chemical composition and microbial flora of seafood vary considerably between species, different fishing grounds and seasons.

Both salt water and fresh water fish contain comparatively high levels of proteins (about 18 per cent) and other nitrogenous constituents. Non-fatty fish such as cod and haddock have a lipid content of less than 1 per cent in contrast to fatty fish such as herring and mackerel which can have lipid contents of up to 30 per cent. The spoilage of salt and fresh water fish appears to occur in essentially the same manner.

It is generally accepted that the internal flesh of healthy, live fish is sterile; microorganisms that exist on fresh fish are generally found in the gills, the outer slime, and the intestines. The *post mortem* changes leading to spoilage depend principally on the chemical composition of the fish, its microbial flora and subsequent handling, processing and storage.

Immediately *post mortem* a whole series of tissue enzyme reactions begin the process of autolysis (basically self digestion of the fish muscle) which leads eventually to spoilage. The autolytic enzyme reactions predominate for 4–6 days at 0°C after which the products of bacterial activity become increasingly evident with the appearance of undesirable odors and flavors. The rate of the autolytic changes are determined by many factors with the most important being temperature, pH, availability of O_2 and the physiological condition of the fish before death.

As spoilage proceeds there is a gradual invasion of the flesh by bacteria from the outer surfaces. Because bacteria can generally only use very basic nutrients as food, bacterial spoilage does not usually commence in whole fish until autolysis is well advanced. Breakdown of the muscle structure only occurs after spoilage has proceeded well beyond the point of rejection. The development of objectionable slimes, odors and flavors results mainly from bacterial activity.

The major spoilage organisms found on spoiled fish include *Pseudomonas, Moraxella, Acinetobacter, Flavobacterium,* and *Cytophaga* species. From a microbiological safety standpoint, the organisms of greatest concern when dealing with MAP fish products are the nonproteolytic *C. botulinum* type E.

Chapter 15

Sterilization of Packaging Material

15.1 Introduction

The required count reduction (number of D values) for the sterilization of the food contact packaging material surface is determined by the type of product, its desired shelf life and likely storage temperature. For non-sterile acidic products of pH <4.5, a minimum of four decimal reductions (4D's) in bacterial spores is required. For sterile, neutral low acid products of pH >4.5, a six decimal reduction is required. However if there is the possibility that *Clostridium botulinum* is able to grow in the product, then a full 12D process should be given.

The non-sterility rate or error rate E_r is the number of non-sterile or faulty packages as a proportion of the total number of packages processed over a given period. It can be calculated using the following equation:

$$E_r = \frac{N}{R} A$$

where N is the number of microorganisms on, and A the area of, the food contact surface of the packaging material and R is the count or number of decimal reductions obtained in the sterilization process. Smaller containers with a smaller food contact surface area will have correspondingly less initial contamination and a less severe sterilization process will be needed to give a certain non-sterility rate E. Conversely, larger containers will require a more severe sterilization process. However, since container volume varies with the cube of the linear dimensions while surface area varies only with the square of the dimensions, the variation in the sterilization process with container size is less than might be expected.

It has been suggested that only 3 per cent of the total number of microorganisms on the package surface are spores. An upper value of 1000 microorgarlisms m^{-2} (30 spores m^{-2}) has been assumed for plastic films and paperboard laminates on reels, and 3000 microorganisms m^{-2} (90 spores m^{-2}) for prefabricated cups. Using these values, the non-sterility rate for a variety of different package sizes and types has been calculated and is reproduced in Table 15.1.

Table 15.1: Non-sterility Rate E$_r$ for Different Package Types, Package Sizes and Count Reductions D

	From the Reel				Prefabricated
	Carton Package			Plastic Cups	Plastic Cups
	200 mL	500 mL	1 L	15 mL	150 mL
Spore Count N	30	30	30	30	90
Surface Area A	0.025	0.045	0.075	0.0037	0.018
R = 10^3					1.6 × 10^{-3}
R = 10^4	0.75 × 10^{-4}	1.4 × 10^{-4}	2.3 × 10^{-4}	1 × 10^{-5}	1.6 × 10^{-4}
R = 10^5	0.75 × 10^{-5}	1.4 × 10^{-5}	2.3 × 10^{-5}	1 × 10^{-6}	1.6 × 10^{-5}
R = 10^6	0.75 × 10^{-6}	1.4 × 10^{-6}	2.3 × 10^{-6}	1 × 10^{-7}	1.6 × 10^{-6}
R = 10^7	0.75 × 10^{-7}	1.4 × 10^{-7}	2.3 × 10^{-7}	1 × 10^{-8}	1.6 × 10^{-7}

In order to attain a non-sterility rate of 10^{-6} in a 1 liter paperboard package of UHT milk, the sterilization process would have to result in 6.4D; for smaller packages, 6 decimal reductions would suffice. Table 15.1 indicates that the smaller the package and therefore the smaller the food contact surface, the lower the non-sterility rate E_r resulting from the packaging material sterilization process.

Because of the importance of the initial level of microbial contamination of the packaging material, steps should be taken to ensure that it is as low as possible. Thus it should be produced, transported and stored under conditions which are as free from microorganisms as possible.

Three main sterilization processes for packaging material are in common use, either individually or in combination: irradiation, heat and chemical treatments. These will each be considered in turn.

Aseptic filling systems provided that the irradiated materials are smooth, UV-resistant and free from dust particles to avoid a "shadowing" effect on the surface. As well, it is important that the irradiation intensity is uniform and adequate for sterilization over the whole of a container which may be of complex shape. It is generally only used commercially in combination with hydrogen peroxide.

Infra-red Radiation

Infra-red (IR) rays are converted into sensible heat when they contact an absorbent surface, resulting in an increase in the temperature of the surface. As with UV irradiation, IR irradiation is only applicable on smooth, even surfaces. IR rays have been used to treat the interior of aluminum lids which have been coated with a plastic lacquer material. Because of the possible softening of the plastic lacquer, the maximum temperature must not exceed 140°C.

Ionizing Radiation

Particle irradiation techniques using gamma rays from cobalt 60 or cesium 139 have been used to sterilize the interior of sealed but empty containers, especially those made of materials which cannot withstand the temperatures needed for thermal sterilization, or which, because of their shape, could not be conveniently sterilized by other means. The bags made from plastic laminates (typically having a molded plastic filling connection) for use in aseptic bag-in-box systems are a good example of the latter category. They are given a radiation dose of 25 kGy (2.5 Mrad) or more which is sufficient to ensure sterility. They are sealed into microbial-proof containers prior to the irradiation treatment. A 20 kGy dose of high-energy electrons was found to sterilize 9 mm of a polyethylene strip infected with approximately 10^5 spores of *B. stearothennophilus*.

It is also possible to use low-energy (100 keV), large area electron beams for the surface sterilization of packaging materials and containers.

Saturated Steam

The most reliable sterilant is without doubt wet heat in the form of saturated steam. However, there are three major problems in the use of saturated steam. The first is that, in order to reach temperatures sufficiently high to achieve sterilization in a few seconds, the steam (and thus, the packaging material with which it comes into contact) must be under pressure, necessitating the use of a pressure chamber. Secondly, any air which enters the pressure chamber with the packaging material must be removed otherwise it would interfere with the transfer of heat from the steam to the package surface, Thirdly, condensation of steam during heating of the packaging material surface produces condensate which may remain in the container and dilute the product.

Despite these problems, saturated steam under pressure is being used to sterilize plastic containers. For example, immediately after deep-drawing, molded plastic cups of polystyrene and the lid foil are subjected to steam at 165°C and 600 kPa for 1.4 s (cups) and 1.8 s (lids). In order to limit the heating effect to the internal surface of the cups, the exterior of the cups is simultaneously cooled. This process has been shown to achieve a 5–6D reduction in *B. subtilis* spores.

Superheated Steam

Superheated steam was the method used in the 1950s for the sterilization of tinplate and aluminum cans and lids in the Martin-Dole aseptic canning process. The cans are passed continuously at normal pressure under saturated steam at 220–226°C for times of 36–45 s depending on the material of construction, aluminum cans having a shorter heating time because of their higher thermal conductivity.

Hot Air

Dry heat in the form of hot air has the advantage that high temperatures can be reached at atmospheric pressure, thus simplifying the mechanical design problems for a container sterilization system.

Hot Air and Steam

A mixture of hot air and steam has been used to sterilize the inner surfaces of cups and lids made from polypropylene which is thermally stable up to 160°C. In this process, hot air is blown into the cups through a nozzle in such a way that the base and walls of the cup are uniformly heated.

Extrusion

During the extrusion of plastic granules prior to blow molding of plastic containers, temperatures of 180–230°C are reached for up to 3 min. However, because the temperature distribution inside the extruder is not uniform and the residence time of the plastic granules varies considerably, it is not possible to guarantee that all particles will achieve the minimum temperature and residence time necessary to result in sterility.

15.2 The Cup Systems Used in Aseptic Filling

Performed Plastic Cups

This is one of the most common types of container for aseptic filling. Although the cups are usually made from high impact polystyrene (HIPS) or polypropylene lit, they can be made from coextruded, multilayered material if improved barrier properties are required. A typical example of the latter cup would be an outer layer of HIPS, then a laminating adhesive; a barrier layer of PVC/PVdC or EVOH copolymer; a laminating adhesive; and finally LDPE. Such a construction with a PVC/PVdC copolymer barrier layer is claimed to give a non-refrigerated shelf life for pudding of more than twelve months.

The cups are fed onto a conveyor which is inside a sterile tunnel supplied with sterile air. The cups are sprayed inside with 35 per cent H_2O_2 solution (35 mg for a 250 mL cup size) and after about 3 s the solution is removed with compressed hot air at a maximum temperature of about 70°C, depending on the material from which the cups are made. The inside surface of the cups reaches a temperature of about 70°C which completes surface sterilization and reduces the peroxide residues to acceptable levels.

Cups can also be sterilized by carrying them through a 35 per cent peroxide bath at 85–90°C before heating and passing through a water bath. The cups then enter a sterile chamber where sterilization is completed by spraying with sterile water and drying with hot air.

The sealing material (usually aluminum foil with a thin coating of a thermoplastic to provide sealability) is typically sterilized with a 35 per cent peroxide solution which is then removed either by radiant heat, hot sterile air or by passing the material over a heated roller. In some systems UV radiation is used, either alone or in conjunction with peroxide.

Form-fill-seal Cups

The plastic material (commonly polystyrene because it is easily thermoformed) in the form of a web is fed from a roll into a thermoformer to give multiple containers still in web form. More complex coextruded multilayer materials incorporating a barrier layer of either PVC/PVdC or EVOH copolymer can also be treated in this way. If an aluminum foil layer is incorporated into the laminate, mechanical forming rather than thermoforming is used.

The advantages in thermoforming cups from a reel compared to using premade cups include a favorable price ratio; simplified

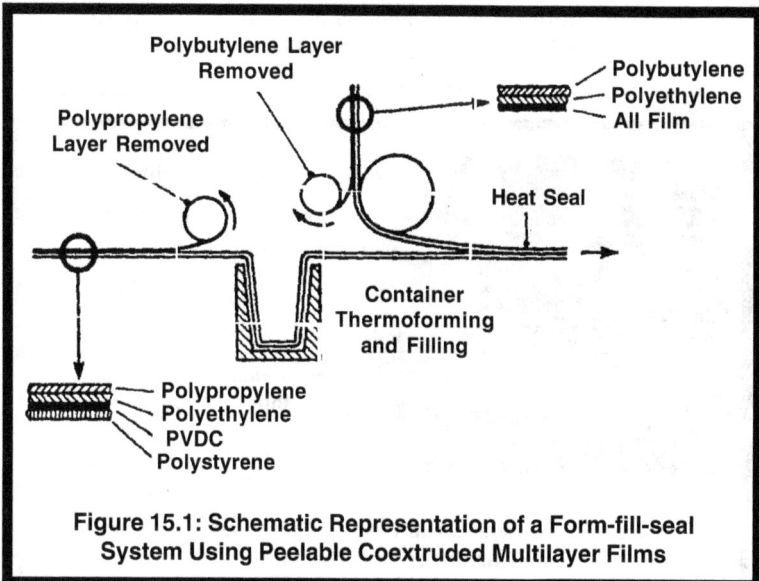

Figure 15.1: Schematic Representation of a Form-fill-seal System Using Peelable Coextruded Multilayer Films

handling since the constant reloading of magazines is avoided; higher output by utilizing multiple tools; lower storage volume requirements for the packaging materials; and maximum sterility of the cups (both inside and outside surfaces) as well as the lidding material by running flat material through sterilizing baths.

15.3 The Carton Material

The carton material normally consists of layers of bleached and unbleached paperboard coated internally and externally with polyethylene, resulting in a carton which is impermeable to liquids and in which the internal and external surfaces may be heat sealed. There is also a thin layer of aluminum foil which acts as an oxygen barrier. The structure of a typical paperboard carton is shown in Figure 15.2. The functions of the various layers have been described as follows:

1. The outer polyethylene protects the ink layer and enables the package flaps to be sealed;

2. The bleached paperboard serves as a carrier of the decor and gives the package the required mechanical rigidity;

3. The laminated polyethylene binds the aluminum to the paperboard;

4. The aluminum foil acts as a gas barrier and provides protection of the product from light;

5. The two inner polyethylene layers provide a liquid barrier.

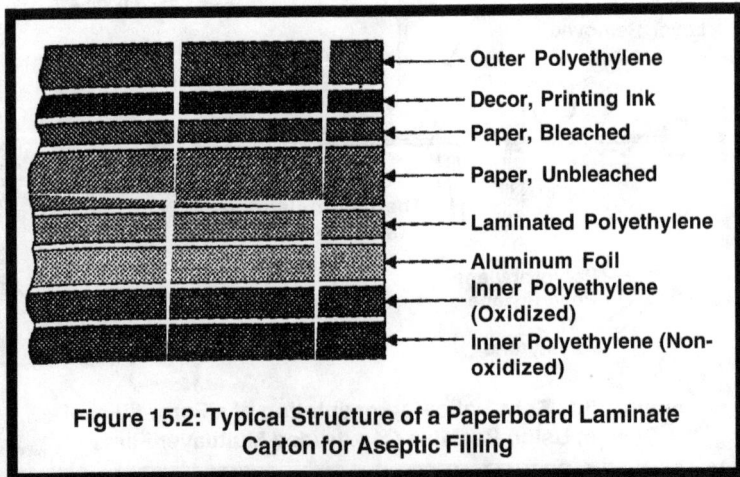

Figure 15.2: Typical Structure of a Paperboard Laminate Carton for Aseptic Filling

Layers labeled:
- Outer Polyethylene
- Decor, Printing Ink
- Paper, Bleached
- Paper, Unbleached
- Laminated Polyethylene
- Aluminum Foil
- Inner Polyethylene (Oxidized)
- Inner Polyethylene (Non-oxidized)

Form-fill-seat Cartons

The packaging material is supplied in rolls which have been printed and creased, the latter being necessary to ease the forming process. A polyethylene strip is sealed to one edge (the reason for this is described later) and the packaging material sterilized using one of two procedures: a wetting system or a deep bath system.

In the wetting system, a thin H_2O_2 film (15–35 per cent concentration) containing a wetting agent to improve the formation of a liquid film is applied on the inner packaging material surface. The material then passes through a pair of rollers to remove excess liquid and is formed and sealed into a cylinder. It then passes under a tubular electric heater which heats the inside surface to about 120°C and evaporates the H_2O_2.

In the deep bath system the packaging material is fed through a deep bath containing H_2O_2 (35 per cent concentration) at 78°C, the residence time being 6 s. After squeezer rollers have removed much of the peroxide, air at 125°C is directed through nozzles onto both sides of the material to heat it to increase the sterilizing effect and to evaporate the peroxide.

The sterilized packaging material is fed into a machine where it is formed into a tube and closed at the longitudinal seal by a heat sealing element. In the process the strip of polyethylene which was added prior to sterilization is heat sealed across the inner surface of the longitudinal seal. It provides protection of the aluminum and paperboard layers from the product which could corrode or swell the layers if such a strip were absent. Product is then filled into the tube and a transverse seal made below the level of the product, thus ensuring that the package is completely filled. Alternatively, the packages may be produced with a headspace of up to 30 per cent of the total filling volume by the injection of either sterile air or other inert gases. The sterilization, filling and sealing processes are all performed inside a chamber maintained at an over-pressure of 0.5 atmos with sterile air.

The sealed packages are then pressed by molds into rectangular blocks, after which the top and bottom flaps or wings are folded down and heat sealed to the body of the package using electrically heated air.

Prefabricated Cartons

In systems of this type, prefabricated carton blanks are used, the cartons being die-cut, creased and the longitudinal seam completed at the factory of origin. The cartons are delivered to the processor in lay-flat form ready to be finally shaped in the filler and the top and bottom seams formed and bonded. Advantages cited for prefabrication include that the fact that the critical longitudinal seam can be flame welded (thus, ensuring good integrity of the seal), and that the same machine can be use used to produce different capacity cartons (provided that the same cross section is used) simply by adjusting the height.

The partly assembled lay-flat blanks are shipped in boxes to the processor, ready for loading into the filler. Stacks of blanks are loaded into a magazine from which they are individually, removed by suction pads opened up into a rectangle and placed on a mandrel. The polyethylene sealing surface at the bottom of the carton is softened by blasts of hot air. As the mandrel wheel turns on towards the bottom pressing station, the bottom is folded by means of rotary transverse and longitudinal folders. The bottom is then pressed and sealed against the end face of the mandrel, after which the carton is transferred to a pocket on the conveyor where the top is prefolded. Although all of the above operations take place under nonsterile conditions, steps are taken to avoid recontamination.

The carton then passes to the aseptic area which consists of several separate functional zones where operations are carried out in sequence. Sterility is maintained in each zone by a slight overpressure of sterile air. The inside surface of the carton is sterilized with a 35 per cent solution of H_2O_2, delivered either as a fine spray or as peroxide vapor in hot air so that the vapor condenses as liquid peroxide on the carton surface. The peroxide is removed by a jet of hot air at 170–200°C. Alternatively, the inside of the carton can be sprayed uniformly with a 1–2 per cent solution of H_2O_2 and then irradiated for about 10 s with high intensity UV radiation. The peroxide is then heated and removed by hot-air jets. Because the total quantity of peroxide used in this latter process is 20–30 times less than in the former, the problems of residual peroxide in the carton and peroxide contamination in the surrounding atmosphere are more easily dealt with.

The next stage of the process is filling. A certain amount of headspace is always advisable to ensure that the package can be opened and poured without spilling. When the contents require shaking (as is the case with flavored milk drinks and pulpy fruit juices), a headspace is essential. For products such as fruit juice it is advantageous to fill the area between the product and the top of the package with steam or an inert gas such as nitrogen. If steam is used, the headspace volume is reduced as a result of a condensation of the steam. A headspace is also crucial when it is not possible to seal the filled package through the product such as when it contains particulate solids. A headspace ensures that sealing can occur above the product line, thus preventing solid material from getting caught up in the top seam where its presence would lead to loss of sterility.

After filling, the carton top is folded and closed, the seal being made either by heating or ultrasonic welding. Production and date codes are added afterwards by ink jet printing or burning into the top seam. Although conventional cartons are of the "gable-top" type, cartons of the brick type can also be produced using basically similar equipment, with the additional step that the gables resulting from the top seam sealing are formed into a flat top. The protruding flaps or "ears" on each side are folded down and sealed to the package with hot air. The finished cartons are then discharged from the pocket chain onto a conveyor belt, ready for the final packaging process.

Chapter 16

Shelf Life of
Packaged Food

16.1 Quality of Food

The quality of foods and beverages decreases with storage or holding time.

Exceptions include distilled spirits (particularly whiskeys and brandies) which develop desirable flavor components during storage in wooden barrels, some wines which undergo increases in flavor complexity during storage in glass bottles, and many cheese varieties where enzymic degradation of proteins and carbohydrates, together with hydrolysis of fat and secondary chemical reactions lead to desirable flavors and textures in the aged cheeses.

However, for the majority of foods and beverages in which the quality decreases with time, it follows that there will be a finite length of time before the product becomes unacceptable. This time from production to unacceptability is referred to as shelf life.

There is no simple, generally accepted definition of shelf life in the food technology literature. The National Food Processors Association in the USA recommended in 1978 as a working professional definition for internal industry use the following: "A product is within its shelf life when it is neither misbranded nor adulterated; when the product quality is generally accepted for its purported use by a consumer; and so long as the container retains its integrity with respect to leakage and protection of the contents."

The Institute of Food Technologists in the USA has defined shelf life as the period between the manufacture and the retail purchase of a food product, during which time the product is in a state of satisfactory quality in terms of nutritional value, taste, texture and appearance. An alternative definition is that shelf life is the duration of that period between the packing of a product and its use, for which the quality of the product remains acceptable to the product user.

The first and last definitions are to be preferred, since the consumer can generally only judge the quality of the product at the time of consumption, not at the time of purchase. Thus, the shelf life refers to the time for which a food can remain on both the retailer's and consumer's shelf before it becomes unacceptable.

Although the shelf lives of foods vary, they are generally determined routinely for each particular product by the manufacturer or processor. Storage studies are an essential part of food product development, with the manufacturer attempting to provide the longest shelf life practicable consistent with costs and the pattern of handling and use by distributors, retailers and consumers.

Inadequate shelf life will often lead to consumer dissatisfaction and complaints. At best, such dissatisfaction will eventually affect the acceptance and sales of brand name products, while at worst it can lead to malnutrition or even illness. Therefore, food processors give considerable attention to determining the shelf lives of their products.

It has been standard practice in almost all food manufacturing and processing establishments to put a "closed" code (so called because only those with knowledge of the coding system can interpret the code) onto the packaged product, sometimes on the primary package, but frequently on the secondary package. Generally this code indicates the time of processing and packaging, *e.g.*, day and year, or shift, day and year, or hour, day and year. For many canned foods, it has long been mandatory to include such information on the end of the container in an embossed form, together with a code for the product itself to aid identification of unlabeled cans in the factory.

Since the advent of the consumer movement in the early 1970s, many different types of open dating systems have been proposed as part of the consumer's "right to know." An open date on a food product is a legible, easily read date which is displayed on the package with the purpose of informing the consumer about the shelf life of the product. Several types of dates can be used.

Pack Date: The date on which the product was packed into its primary package, *i.e.*, the immediate container in which it is sold. It does not provide any specific information as to the quality of the product when purchased or how long it will retain its quality after purchase.

Display Date: The date on which the product was placed on the shelf by the retailer.

Pull Date or Sell by Date: The last date on which the product should be sold in order to allow the consumer a reasonable length of time in which to use it.

Best Before or Best if Used by Date: The last date of maximum high quality.

Use by Date or Expiration Date: The date after which the food is no longer at an acceptable level of quality.

This form of dating is used infrequently because quality changes generally occur slowly and it is simply not possible to state that a food will be acceptable one day and unacceptable the next. An exception is with baked goods such as bread.

There is no uniform or universally accepted open dating system used for food throughout the world. In some countries, mandatory open dating of all perishable and sometimes semiperishable foods is required, while in other countries such requirements are voluntary. Arguments can be advanced both for and against the open dating of foods. However, there has been an obvious increase in the quantity of open dated food on sale throughout the world, and this trend is likely to continue.

The European Community (EC) has adopted a "minimum durability date". This is the length of time for which the product may be stored (the storage conditions may be specified, *e.g.*, temperature not to exceed 7°C) before its quality has deteriorated to such an extent that it would be regarded as unacceptable by the manufacturer or the consumer. This concept (basically equivalent to the "best before" date defined above) allows the processor to set the quality standard of the food, since the product would still be acceptable to many consumers after the "best before" date had passed.

16.2 The Alteration of Shelf Life

The alteration of product shelf life is controlled by following factors:

1. Product characteristics;
2. The environment to which the product is exposed during distribution; and
3. The properties of the package.

The shelf life of a product will be altered by changing its composition and form, the environment to which it is exposed, or the packaging system.

Based on the nature of the changes which can occur during storage, foods may be divided into three categories–perishable, semi-perishable and non-perishable or shelf stable, which translate into very short shelf life products, short to medium shelf life products, and medium to long shelf life products.

Perishable foods are those that must be held at chill or freezer temperatures (*i.e.*, 0 to 7°C or –12 to –18°C, respectively) if they are to be kept for more than short periods. Examples of such foods would include milk, fresh flesh foods such as meat, poultry and fish, and many fresh fruit and vegetables.

Semi-perishable foods are those which contain natural inhibitors (*e.g.*, some cheeses, root vegetables and eggs) or those that have received some type of mild preservation treatment (*e.g.*, pasteurization of milk, smoking of hams and pickling of vegetables) which produces greater tolerance to environmental conditions and abuse during distribution and handling.

Shelf stable foods are considered "non-perishable" at room temperatures. Many unprocessed foods fall into this category, and are unaffected by microorganisms because of their low moisture content (*e.g.*, cereal grains and nuts, and some confectionery products). Processed food products can be shelf stable if they are preserved by heat sterilization (*e.g.*, canned foods), contain preservatives (*e.g.*, soft drinks), are formulated as dry mixes (*e.g.*, cake mixes) or processed to reduce their water content (*e.g.*, raisins or crackers). However, shelf stable foods only retain this status if the integrity of the package which contains them remains intact. Even then, their shelf life is finite due to deteriorative chemical reactions which proceed at room temperature independently of the nature of the packaging, and the permeation through high barrier polymer packages of gases, odors and water vapor.

Bulk Density

The free space volume of a package (V) is directly related to the bulk density (p_b) and the true density (p_p) of the product as follows:

$$V = V_t - V_p = \frac{W}{p_b} - \frac{W}{p_p}$$

where, V_t = total volume of the package; V_p = volume of the product, W = weight of the product.

Thus, for packages of similar shape, equal weights of products of different bulk densities will have different free space volumes, and as a consequence, package areas and package behaviour will differ. This has important implications when changes are made in package size for the same product, or alterations are made to the process, resulting in changes to the product bulk density.

While the true density of a food depends largely on its composition and cannot be changed significantly, the bulk density of food powders can be affected by processing and packaging. Some food powders (*e.g.*, milk and coffee) are instantized by treating individual particles so that they form free-flowing agglomerates or aggregates in which there are relatively few points of contact; the surface of each particle is thus more easily wetted when the powder is rehydrated. Instantization results in a reduction of bulk density, *e.g.*, for skim milk powder from 0.64 to 0.55 g mL^{-1}. A wide range of bulk densities is encountered in food products, from around 0.056 g mL^{-1} for potato chips to 0.96 for granulated salt.

The free space volume has an important influence on the rate of oxidation of foods, for if a food is packaged in air, a large free space volume is undesirable since it constitutes a large oxygen reservoir. Conversely, if the product is packaged in an inert gas, a large free space volume acts as a large "sink" to minimize the effects of oxygen transferred through the film. It follows that a large package area and a low bulk density result in greater oxygen transmission.

In many foods such as those containing whole tissue components, or where the reacting species are partially bound as in membranes, structural proteins or carbohydrates, the concentration varies from one point to another, even at zero time. In other words, there are initial inhomogeneities and discontinuities for each compound. Furthermore, since most of these compounds will have little opportunity to move, the concentration differences will get greater as the reactions proceed out from isolated initial foci. This has been described as the "brush-fire" effect and is especially important in chain reactions such as oxidation.

As well, there may be several different stages of the deteriorative reaction proceeding at once, and the different stages may have different dependence on concentration and temperature, giving disguised kinetics. Such a situation is frequently the case for chain reactions and microbial growth which have both a lag and a log phase with very different rate constants.

The point to be taken from this is that for many foods it may be difficult to obtain kinetic data of use for predictive purposes. In such situations, use of sensory panels to determine the acceptability of the food is the recommended procedure.

16.3 The Transfer of Mass and Heat through the Package

The deterioration in product quality of packaged foods is often closely related to the transfer of mass and heat through the package. Packaged foods may lose or gain moisture; they will also reflect the temperature of their environment, since very few food packages are good insulators. Thus, the climatic conditions (*i.e.*, temperature and humidity) of the distribution environment have an important influence on the rate of deterioration of packaged foods.

In the case of mass transfer, it is the exchange of vapors and gases with the surrounding atmosphere which is of primary concern. Moisture vapor and oxygen are generally of most importance, although it may be necessary to consider the exchange of volatile aromas from the product or to the product from the surroundings. As well as oxygen, transmission of nitrogen and carbon dioxide may have to be taken into account in packages where the concentration of these gases inside the package has been modified from ambient to inhibit or slow down deteriorative reactions in the food.

Generally, the difference in partial pressure of the vapor or gas across the package barrier will control the rate and extent of permeation, although transfer can also occur due to the presence of pinholes in the material, channels in seals and closures, or cracks which result from flexing of the packaging material during filling and subsequent handling. Since the gaseous composition of the earth's atmosphere is essentially constant at sea level, the partial pressure difference of gases across the package barrier will depend on the internal atmosphere of the package at the time the package was sealed.

In contrast to the common gases, the partial pressure of water vapor in the atmosphere varies continuously, although the variation is generally much less in controlled climate stores. The magnitudes of these variations in external climate are detailed in, for example, British Standard BS 4672, and in various publications of the U.S. Army Natick Laboratories.

The general pattern is for temperature to rise and relative humidity to fall towards midday, and for relative humidity to rise and temperature to fall through the afternoon and evening. Obviously, this pattern would be modified by rainfall and vary with the seasons. The detailed climatic statistics of global maximum and mean temperatures which are available in many countries are of great assistance. Despite meteorological and secular trends, the daily and even the annual cycle of temperatures can be normalized to a standard cycle with a standard frequency distribution derived from the mean and range at many places. This is because of the sinusoidal trend of diurnal (earth's rotation) and seasonal (earth's revolution) solar radiation intensity. To summarize, mass transfer depends on the partial pressure difference across the package barrier, and on the nature of the barrier itself.

One of the major determinants of product shelf life is the temperature to which the product is exposed during its life time from production to consumption. Without exception, food products are exposed to fluctuating temperature environments during this time, and it is important (if an accurate estimate of shelf life is to be made) that the nature and extent of these temperature fluctuations is known. There is little point in carefully controlling the processing conditions inside the factory and then releasing the product into the distribution and retail system without some knowledge of the conditions which it will experience in that system. Such knowledge is essential in the case of products containing a "best before" or "use by" date.

The storage climates inside buildings such as warehouses and supermarkets are only broadly related to the external climate as reported by weather stations, and climatic variations in temperature and humidity can differ as much between different building constructions as between seasons on one site. Data for temperature in six major cities in the United States has also been presented.

If the major deteriorative reaction causing end of shelf life is known, simple, expressions can be derived to predict the extent of deterioration as a function of available time-temperature storage conditions. The basic types of deteriorative reactions which foods undergo were discussed later, together with the rates of these reactions and the factors controlling these rates. These reactions and their rates will now be analyzed in relation to product shelf life.

Fundamental to such an analysis is that the particular food under consideration follows the "laws" of "additivity" and "commutativity". Additivity implies that the total extent of the degradation reaction in the food produced by a succession of exposures at various temperatures is the simple sum of the separate amounts of degradation, regardless of the number or spacing of each time-temperature combination. Commutativity means that the total extent of the degradation reaction in the food is independent of the order of presentation of the various time-temperature experiences.

16.4 Shelf Life Testing of Food Products

The shelf life testing of food products can be studied under three categroies:

1. Experiments designed to determine the shelf life of existing products,

2. Experiments designed to study the effect of specific factors and combinations of factors such as storage temperature, packaging materials or food additives on product shelf life; and

3. Tests designed to determine the shelf life of prototype or newly developed products.

There are several basic approaches to determine the shelf life of a food product:

Literature Study

The shelf life of an analogous product is obtained from the published literature or in-house company files.

Turnover Time

The average length of time which a product spends on the retail shelf is found by monitoring sales from retail outlets, and from this the required shelf life is estimated. This does not give the "true' shelf life of the product but rather the "required" shelf life, it being implicitly assumed that the product is still acceptable for some time after the average period on the retail shelf.

End Point Study

Random samples of the product are purchased from retail outlets and then tested in the laboratory to determine their quality; from this a reasonable estimate of shelf life can be obtained since the product has been exposed to actual environmental stresses encountered during warehousing and retailing.

Accelerated Shelf Life Testing

Laboratory studies are undertaken during which environmental conditions are accelerated by a known factor so that the product deteriorates at a faster than normal rate. This method requires that the effect of environmental conditions on product shelf life can be quantified.

Regardless of the method chosen or the reasons for its choice, sensory evaluation of the product is likely to be used either alone or in combination with instrumental or chemical analyses to determine the quality of the product. Not surprisingly, many food scientists and technologists in industry attempt to replace human judgment with instrumental or chemical analyses since the latter are not prone to fatigue or subject to the physiological and psychological fluctuations which characterize human performance. However, since it is human judgment which is the ultimate arbiter of food acceptability, it is essential that the results obtained from any instrumental or chemical analysis correlate closely with the sensory judgments for which they are to substitute.

That such is not always the case has been well documented. In chemical and physical tests, analytical parameters are isolated so that a single signal is monitored, whereas sensory responses are more complex due to the integration of multiple signals due to the interdependence of appearance, texture, aroma and flavor of a food. Hedonic (*i.e.,* like/dislike) responses, and to a lesser extent intensity (*i.e.,* stimulus strength such as firm/very firm) judgments are subject to many experimental influences such as past exposure to the product and those created by the actual test protocol (*e.g.,* number of samples and order of presentation). The potential fallacy of correlating hedonic responses with physical and chemical measurements has been discussed using as examples the color, mouth feel and sweetness of specific foods.

Different methods are used to measure whether "fresh" samples are different from stored samples, or "control" samples from test samples. These methods require trained or experienced panelists, and three experimental designs are commonly used for the purpose of shelf life testing the paired comparison test; the duo-trio test, and the triangle test. Further details about these tests can be found in standard texts on sensory evaluation. Descriptive methods are used to measure quantitative and/or qualitative characteristics of products and require specially trained panelists. Effective methods are used to evaluate preference, acceptance and/or opinions of products and do not require trained panelists.

The selection of a particular sensory evaluation procedure for evaluating products undergoing shelf life testing is dependent on the purpose of the test. Acceptability assessments by untrained panelists are essential to an open-dating program, while discrimination testing with trained panelists might be used to determine the effect of a new packaging material on product stability.

For shelf life determinations for new products and in situations where freshness standards cannot be maintained for the duration of the storage period, descriptive profile analysis is often employed.

In shelf life testing there can be one or more criteria which constitute sample failure. One criterion is an increase or decrease by a specified amount in the mean panel score. Another criterion is microbial deterioration of the sample to an extent which renders it unsuitable or unsafe for human consumption. Finally, any physical changes such as changes in odor, color, texture, flavor, etc. that render the sample unacceptable to either the panel or the consumer are criteria for product failure. Thus sample failure can be defined as the condition of the product which exhibits either physical, chemical, microbiological or sensory characteristics that are unacceptable to the consumer, and the time required for the product to exhibit such conditions is the shelf life of the product.

However, a fundamental requirement in the analysis of data is knowledge of the statistical distribution of the observations, so that the mean time to failure and its standard deviation can be accurately estimated, and the probability of future failures predicted. As has been pointed out, the length of shelf life for food products is usually obtained from simple averages of time to failure on the assumption that the failure distribution is symmetrical. If the distribution is skewed, estimates of the mean time to failure and its standard deviation will be biased. Furthermore, when the experiment is terminated before all the samples have failed, the mean time to failure based on simple averages will be biased because of the inclusion of unfailed data.

In order to improve the method of estimation of shelf life, knowledge of the statistical distribution of shelf life failures is required, together with an appropriate model for data analysis. Five statistical models normal, lognormal, exponential, Weibull and extreme-value distributions were fitted to failure data using the method of hazard plotting which provided information about the adequacy of fit of the observed data to the proposed model, the mean or median time to failure, and the probability of future failures. The Weibull distribution was suggested as the most appropriate shelf life model, and examples of its use to predict end of shelf life for food products have been presented.

The nature of product failure over time is commonly represented by a 'bathtub' curve which has many applications in the actuarial and engineering sciences.

During the time between X_0 and X_1, early failures may occur due to faulty packaging (*e.g.*, pinholes or poor seals) and product abuse. However, the early failures should not be taken as true failures relative to the shelf life of the product. From X_1 to X_i, no product failures (barring random fluctuation) would be expected. From the time Xi to the termination time X_n, the hazard (failure) rate increases, this time representing the true failure due to deteriorative changes within the product. The length of shelf life is determined between the times X_0 and X_n, and the hazard function plays a central role in the analysis of failure data.

One problem with shelf life testing is to develop experimental designs which minimize the number of samples required, thus

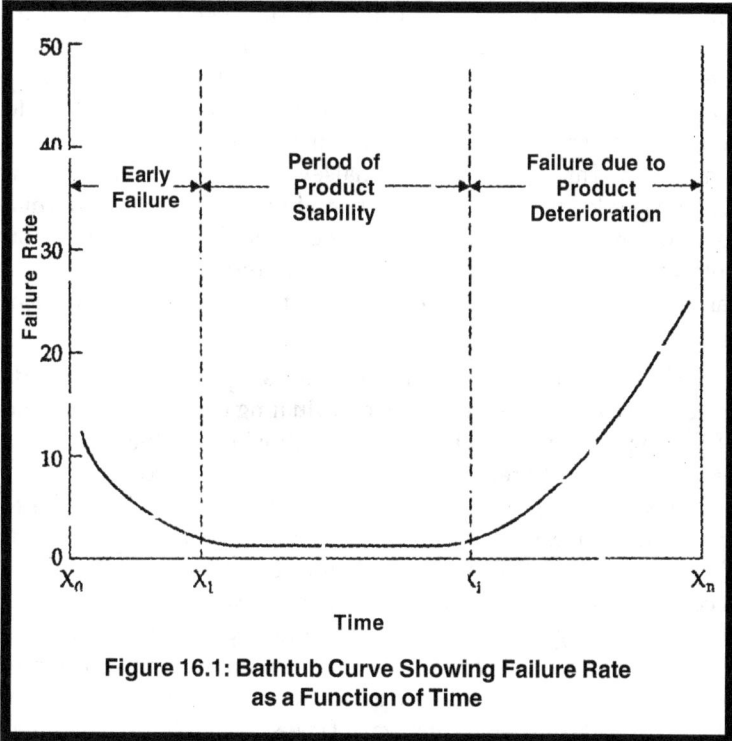

Figure 16.1: Bathtub Curve Showing Failure Rate as a Function of Time

minimizing the cost of the testing while still providing reliable and statistically valid answers. Three experimental designs for use in shelf life testing of foods (partially staggered; staggered, and completely staggered) have been presented by Gacula and the interested reader should consult this paper for further details.

More recently, workers at the U.S. Army Natick R&D Center described a storage study of a military ration (called Meal, Ready-to-Eat) using sensory data, with the purpose of finding the effects of time and temperature on shelf life estimates, taking into account their variance. Four algorithms were used in the calculations: the contingency-table, logit, linear and nonlinear regressions. This paper should be consulted for further details.

16.5 ASLT and Dehydrated Products

The basic assumption underlying accelerated shelf life testing (ASLT) is that the principles of chemical kinetics can be applied to quantify the effects which extrinsic factors such as temperature, humidity, gas atmosphere and light have on the rate of deteriorative reactions. These basic principles and the way in which they can be applied to foods have been described later. By subjecting the food to controlled environments in which one or more of the extrinsic factors is maintained at a higher than normal level, the rates of deterioration will be speeded up or accelerated, resulting in a shorter than normal time for product failure. Because the effects of extrinsic factors on deterioration can be quantified, the magnitude of the acceleration can be calculated and the "true" shelf life of the product under normal conditions calculated.

The need for ASLT of food products is simple: since many foods have shelf lives of at least one year, evaluating the effect on shelf life of a change in the product (*e.g.*, a new antioxidant or thickener), the process (*e.g.*, a different time/temperature sterilization regime) or the packaging (*e.g.*, a new polymeric film) would require shelf life trials lasting at least as long as the required shelf life of the product. Companies cannot afford to wait for such long periods before receiving knowing whether or not the new product/process/packaging will give an adequate shelf life, because other decisions (*e.g.*, to construct a new factory; order new equipment; or arrange contracts for the supply of new packaging material) have led times of months and/or years. Some way of speeding up the time required to determine the shelf life of a product is necessary, and ASLT has been developed for that reason. Such procedures have long been used in the pharmaceutical industry where shelf life and efficacy of drugs are closely related. However, the use of ASLT in the food industry is not as widely used as it might be, due in part to the lack of basic data on the effect of extrinsic factors on the rates of deteriorative reactions, in part to ignorance of the methodology required, and in part to a healthy skepticism of the advantages to be gained from using ASLT procedures.

Dehydrated Products

In dehydrated vegetables, lipid and hydrolytic oxidation, together with nonenzymic browning and (in the case of green vegetables) chlorophyll degradation, are the major modes of deterioration. In dehydrated fruit, the major modes of deterioration is nonenzymic browning. Samaniego used temperatures of 30° and 40°C to accelerate deterioration in sliced green beans and onion flakes and found that, for example, at 40°C, the shelf life of onions was 11 times shorter, and at 30°C, 3.5 times shorter than at 20°C when the a_w was 0.56.

Canned Foods

It is generally assumed that if good manufacturing practices are followed, microbial deterioration of canned foods will not be a problem. Where there is thermophilic spoilage when canned foods are stored at elevated temperatures, this is more than likely due to inadequate cooling of the cans following thermal processing. Microbial spoilage at ambient temperatures is generally the result of 'leaker' spoilage, so called because the microorganisms are drawn into the can during cooling; chlorination of cooling water according to good manufacturing practice will alleviate this problem. Thus deteriorative reactions in canned foods will normally be limited to organoleptic changes such as loss of color, development of undesirable flavors, and nutrient degradation.

Labuza quotes a producer of canned meat products as stating that the major mode of deterioration is hydrogen gas production resulting from internal corrosion of the can. Samples were stored at 37.8°C to accelerate this deterioration; the shelf life at 37.8°C was considered to be 40 per cent of the shelf life at 4.4°C, corresponding to a Q_{10} of 1.3.

Starches

The use of accelerated tests for long shelf life products (*i.e.*, those which retain all their key characteristics after 18 months storage) which contain starches has been described. Such products include sterilized products packaged to exclude recontamination, and products formulated to inhibit spoilage. In these situations, temperature is used as the accelerating factor and a typical scheme. Each product should be tested in the actual package which is to be used to confirm that there are no problems with product/package compatibility.

A product which retains its desirable qualities after being subjected to the tests has a high probability of having a minimum shelf life of 18 months in normal distribution. However, caution is still necessary since interpretation of the results requires experience of similar product systems. This is particularly true of the 55°C tests where acid hydrolysis, color changes, etc. may be severe and yet are not reproduced at ambient temperature even after two years storage.

ASLT methodology for fatty foods has been reviewed with particular emphasis on the testing of antioxidant effectiveness. In all the classical ASLT methods, temperature is the dominant acceleration factor used, and its effect on the rate of lipid oxidation is best analyzed in terms of the overall activation energy E_A for lipid oxidation. An inherent assumption in these tests is that E_A is the same in both the presence and absence of antioxidants, although indications are that it is in fact considerably lower in the latter case.

Other acceleration parameters which are used for shelf life are the oxygen pressure, reactant contact and the addition of catalysts. The effect of these factors is generally much less important than that of temperature. An exception would be high-fat products packaged in metal containers where metal contamination of the product may be the most important factor in limiting shelf life.

16.6 Time-temperature Records of Foods

Time-temperature recorders have been used for many years to provide a record of the temperature history to which a food has been subjected. Until recently they have tended to be relatively expensive, but with the commercialization of micro-chip technology, temperature recorders at quite modest cost are now available. They are in common use to record the temperature of frozen foods and fresh horticultural produce during transportation by sea. When the shipment arrives at a port, the temperature history of the product is inspected. If the product has been subjected to too high a temperature for too long a time, the load may be rejected without necessarily inspecting the product. This action is done on the basis that quality deterioration is cumulative and the product has lost enough of its shelf life to expect consumer dissatisfaction at a later stage. Alternatively, the decision may be made to dispose of the product quickly by discounting its price or some other marketing strategy.

The limitations of time-temperature recorders lie in the fact that they are comparatively expensive, can only monitor the specific area of the load where their sensors (typically thermistors or thermocouples) are located, and cannot follow the product through all phases of the distribution system unless a sensor is placed on each individual case or pallet. A further limitation concerns the interpretation of the data from a recorder; if the product has been subjected to a period at high temperature, how much of its shelf life has it lost? The answer to such a question is not simple, and often requires a value judgement by a qualified food technologist. In practice, such decisions need to be made by agents or brokers who are on the scene when the product arrives so that appropriate action can be taken immediately.

For the above reasons, time-temperature recorders have not, and will not, find widespread use in the food industry for monitoring the shelf life of foods. Instead, the use of time-temperature indicators is likely to become quite widespread over the next few years, due largely to an increasing product quality consciousness and awareness of product liability, especially in the USA.

A time-temperature indicator is a device which responds to the combined effects of time and temperature over a period. Many devices which can be attached to food packages to integrate the time and temperature to which the package is exposed have been developed, and the patent literature contains designs of more than 60 such devices. Initially the majority of these devices were developed specifically for frozen foods, but there is now widespread interest in shelf life devices for most categories of food, especially those foods where the rate of quality deterioration is highly temperature sensitive.

Overviews of the major types of time-temperature indicators have been presented.

There are two classifications of time-temperature indicators devices which respond only after predetermined threshold temperature has been exceeded (these are said to be "partial history" indicators), and devices which respond to all temperature conditions (these are classified as "full history" indicators). Partial history indicators are intended to identify abusive temperature conditions and thus, there is no direct correlation between food quality change and the response of this class of indicator.

Chapter 17

Importance of Eco-friendly Packaging and its Sustainability

In modern societies, on the present level of economic development, to satisfy our needs, we have to rely on the services of the society. To be able to obtain consumer goods, we need among others the service of packaging. In the rapidly urbanising world, packaging is substantial to obtain basic consumer goods. It is also important to satisfy consumers' needs in an effective way. Packaging should fulfil its function with a minimum overall resource use. Packaging is an important part of our lives. This way the most important 'impact' of packaging is that it enables satisfying human needs in an effective way (Eva Pongrácz, 1998).

Packaging effectivity means that packaging fulfils its function with minimal use of resources and minimal overall wastage. Service means that packaging consumer goods, helps their distribution, and gives an access to goods otherwise not accessible. It is most evident in food packaging. Conserving perishable food prevents early spoilage, invention of aseptic packaging prolongs shelf-life and makes possible distribution also to greater distances. This fact is more and more important in the present way of increasing urbanisation. More than 150,000 people are being added to urban population in developing countries every day. Thirty-five years ago only one third of the world's population was urban. The prediction is that by 2025, two-third of the world's people will live in cities. This means that more people will live in cities than occupied the whole planet ten years ago. What makes it worse, by 2015, the world will have 33 'megacities' with populations over 8 million and more than 500 cities with populations of 1 million or more. Greater Tokyo already has 27 million people, Sao Paulo, Brazil 16.4 million; and Bombay, India, 15 million. (World resources, 1996-97). In such level of urbanisation, distribution of goods, especially food, is crucial, and the role of packaging is enormous.

Traditional Food Packaging Technologies

Food Systems Population drift from rural to urban areas has caused drastic changes in the food supply network from farm to the consumer in many emerging nations. One traditional belief which can no longer be sustained is the old saying that, "There are always fish in the rivers and lagoons, and rice, taro, or sweet potato in the fields; therefore let there be no concern for the next meal". It is more likely that the farmer has gone to the city or even overseas, and is earning a laborer's wage to keep his family in food during the off-season. So food is now brought to the market by many and various means, and redistributed to these new consumers. There are new vistas for traditional food markets where the technologies are tested beyond their limits (Hicks, 2002).

Given the circumstances in which many developing countries are today, the challenge for their traditional technologies is that often they do not contribute sufficiently to meeting socio-economic imperatives. This is true also of those food technologies where many of the processing methods have remained unchanged for centuries, and are becoming inadequate to cope with modern needs, because they are too labor-intensive and depend now too much on natural environmental conditions. It is now clear that there is a need to lessen the dependence on nature, reduce the drudgery, shorten the time of the work involved and upgrade the preparation, quality, packaging, presentation, and shelf-life of these traditional foods and their packaging.

The Need for Packaging

It is difficult to answer whether products do need packaging, or the consumer does? For whom is the package made? The product, or the consumer? The package is tailor-made for the product, but one shall not forget that all the products are made for consumers. From the consumer's point of view the package's function is to protect the product. In the case of bulk goods, or if the product cannot be used without the support of the package, the package must help the use of the product. Without the service of packaging, most of the goods, especially food, couldn't reach the consumer. Living in the countryside, one can buy goods, for example milk directly from the farmer, but a can is needed to carry it home. This can also be considered as a refillable package, with almost an unlimited number of uses. In a developed countries, milk can only be bought from the store, packed in aseptic carton box. It is not possible to avoid it, unless one goes daily to the dairy, and drink from the tap.

From the packager's point of view, the most important function of the packaging is the protection of the usage and aesthetic values of the product from damages, and get the product sold to the consumer. For the producer, the package is also a value-creating media of the product. With the help of the package, the product can be sold to the consumer. For example, a barrel full of toothpaste has almost no purchase value, since who wants to buy 100 litre of toothpaste? Most probably nobody would buy even a handful of it in the grocery, but one certainly does buy 50 grams of toothpaste in a tube.

Till the middle of this century, groceries were a meeting point: people talked there. Most of the products were sold in bulk, one could have a look at its quality, discuss with the shopkeeper, asked information about them. The big supermarket stores are meant for shopping. One picks up the goods, pays and leaves. One does not see people talking, or ask somebody: Did you try this product? How is it like? The function of the shopkeeper is taken by the packages, and clearly in the competition of the goods wins that one, which can use the most efficiently those seconds one has for viewing them. The package has to arouse the attention, attract the consumer into arm's-length and influence him or her in buying decision, and in the other hand prepare the next purchase based on the imprinting in the mind.

Modern packagings are an expressive form of the consumer-lifestyle. Over the protective function, packagings are giving character, "personality" to the product. They are following the product to the consumer, giving practical as well as aesthetic value. The proportion of the usage and aesthetic functions is important. The quality of packaging reflects our universal culture. All those products that appear in the shops and offer themselves for purchase, apart of that - with very few exceptions - their usage function is primary, are creating the surrounding material environment.

Minimum Packaging Technology for Processed Foods

Technologies are called traditional if, unaffected by modernization, they have been commonly applied over a long period of time. In general, traditional technologies tend to be cheap, easy to produce, apply, maintain, and repair. They are generally labor-intensive, which can be economically beneficial, but as far as food packaging technologies are concerned, the final products are often hygienically sub-standard and they usually have a short shelf-life.

Many traditional foods have nonetheless remained unchanged in process or package for centuries, due to the fact that they developed in a particular location and are deep rooted in the natural, cultural, religious, and socio-economic environment. Some have disappeared without a trace as a result of modern influences, while others have expanded on a global scale, becoming household products, *e.g.*, soysauce, now a multimillion dollar industry. The reasons for this phenomenon need to be examined (Hicks 1983).

The role of packaging in waste reduction is the most evident at food packaging. When food is processed and packaged, the food residues are often used as fuel, animal feed or some economically useful by-product. In absence of packaged processed food, the residues become garbage in the household. Another reason why food packaging reduces waste is that it reduces spoilage. In developing countries food wastage is between 20-50 per cent because of poor, or the lack of packaging. In Europe, where packaging is used in handling, transport, containment and storage, food wastage is approximately 2-3 per cent. (PIN, 1996) As the use of packaging materials increases, the fraction of food waste decreases. Overall, for every 1 per cent increase of packaging, food waste decreases by about 1.6 per cent. (Scarlett, 1996).

Satisfaction of human needs generally takes place with the use of natural resources. The social development has considerably accelerated in this century. The driving force of this is the demand for goods. Further, the population of the earth is increasing almost exponentially, which multiples the effect of humans on the nature. Neglecting the fact, whether it is possible, if the whole population had the consumption scale of the present maximum level (that of US for example), the effect of it would be catastrophic. (Meadows *et al.*, 1992). It is crucial thus to satisfy our needs in the most optimum way.

Packaging as any other industrial activity, has positive and negative impacts on the environment. The negative impacts include the resource use, and the ecological effects of packaging related wastes and emissions. In a well functioning society, these negative effects shall not be neglected. The effectiveness and efficiency of packaging should be supervised and our consumption habits should be examined.

Food Preservation Principles and their Integration with Food Packaging

Food Unit Operations

The surplus foods grown in the village have a need to be more carefully harvested, protected from spoilage and damage, packaged, and transported by various means to these markets. Unless the goods are sold with minimum spoilage and at their peak flavor, appearance, and nutritional value and presented in an attractive way, they may not be eaten at all. This is a worse situation than if the crop had never been grown and can represent serious loss and waste to a community. Very little investment has been made so far in developing traditional technologies or in applying scientific knowledge in most of the developing countries; meanwhile the more expensive products of imported technologies have further slowed the development of indigenous technologies. It has been increasingly recognized that the time has come when these traditional technologies must be upgraded through scientific application of packaging principles and then integrated with other packaging functions such as marketing and advertisement. In addition, careful environmental considerations need to be given to minimum packaging forms to avoid pollution problems and ensure sustainability (Hicks, 2002).

Before food can be packaged, there are many unit operations involved after harvesting the raw materials, including cleaning, grading, disposal of unwanted material, then stabilization of the enzymatic, biochemical and microbial spoilage. If a study of the preservation and packaging of foods is undertaken, a key question is, "What factors cause spoilage and deterioration in foods". The main factors are microorganisms (bacteria, yeast, and moulds), as well as enzymes, temperature, and biochemical changes in the foods. Food preservation techniques are designed to prevent these spoilage changes and impart a keeping quality or shelf-life to the processed foods.

Packaging is an integral part of the processing and preservation of foods and can influence many of these factors. It can influence physical and chemical changes, including migration of chemicals into foods. The flavor, color, texture as well as moisture and oxygen transfer is influenced by packaging. The effects of temperature changes and light can be modified by packaging materials.

Types of Packaging

Early forms of packaging ranged from gourds to sea shells to animal skin. Later came pottery, cloth and wooden containers. These packages were created to facilitate transportation and trade.

Utilizing modern technology, today's society has created an overwhelming number of new packages containing a multitude of food products. A modern food package has many functions, its main purpose being to physically protect the product during transport. The package also acts as a barrier against potential spoilage agents, which vary with the food product. For example, milk is sensitive to light; therefore, a package that provides a light barrier is necessary. The milk carton is ideal for that. Other foods like potato chips are sensitive to air because the oxygen in the air causes rancidity. The bags containing potato chips are made of materials with oxygen-barrier properties. Practically all foods should be protected from filth, microorganisms, moisture and objectionable odors. We rely on the package to offer that protection.

Aside from protecting the food, the package serves as a vehicle through which the manufacturer can communicate with the consumer. Nutritional information ingredients and often recipes are found on a food label. The package is also utilized as a marketing tool designed to attract the attention at the store. This makes printability an important property of a package.

The food industry utilizes four basic packaging materials: metal, plant matter (paper and wood), glass and plastic. A number of basic packaging materials are often combined to give a suitable package. The fruit drink box is an example where plastic, paper and metal are combined in a laminate to give an ideal package. This concept can be easily seen in your peanut butter jar. The main package containing the food (primary package) is made of glass (or plastic), the lid is made of metal lined with plastic, and the label is made of paper.

Each basic packaging material has advantages and disadvantages. Metal is strong and a good overall barrier, but heavy and prone to corrosion. Paper is economical and has good printing properties; however, it is not strong and it absorbs water. Glass is transparent, which allows the consumer to see the product, but breakable. Plastics are versatile but often expensive. Therefore, combining the basic materials works well in most cases. So, for a product like milk, which is an essential food for children and young adults and therefore cannot be very expensive, paper makes a good economical material. It also provides a good printing surface. However, since paper absorbs water, it will gain moisture from the milk, get weaker and fail, thereby exposing the milk to spoilage factors. It may even break and waste the product. When a thin layer of a plastic called polyethylene is utilized to line the inside of the milk carton, it serves as a barrier to moisture and makes an economical, functional package.

After making a food product and placing it in the appropriate package, a number of these individual packages must be placed in a large container to facilitate shipment. These larger containers are called secondary packages. The paperboard box is a very common secondary package. Plastics also can serve as secondary packages. The milk case in which a number of milk cartons are delivered to the supermarket is a good example.

Driving the Choice: Preservation, Protection and Safety

The main factors that influence the decision of food packaging materials are preservation, protection and safety. Packaging must protect the contents from physical damage and from external contamination *e.g.* microbiological contamination. It must preserve the quality of the contents, whether for short shelf-life of some days or for the extended shelf-life of several months. Food and water packaging must equally protect the nutritional value and taste of the contents.

More importantly, packaging must be safe and have no risk of affecting health. The intrinsically inert nature of polyethylene and polypropylene help make them ideal candidates for safe packaging and they easily comply with relevant national and international regulations.

Upgrading of Food Packaging

Packaging accounts for an increasing share of the costs of the food processing industry, rising from about 4 per cent in 1947 to 10 per cent in 1987, and continuing to rise. Despite new materials which have reduced packaging weight, the total and relative costs of food and beverage packaging are increasing (Connor and Schiek, 1997). On average, the cost of packaging materials represents about one-fifth of material costs, however, in 10 out of 40 food industry sectors, packaging costs exceed the costs of the edible food stuff ingredients (Connor and Schiek, 1997).

In most countries packaging continues to be driven by consumer demand, with regulatory bodies playing a limited role. In Europe, however, legislation and taxes on packaging have been established to encourage reduced packaging waste and more sustainable packaging practices.

Upgrading of traditional food packaging technologies in many cases, introduces exogenous factors, *i.e.*, the importation of technology from abroad. Whether or not adapted to local circumstances, the use of imported packaging technologies in many developing countries remains restricted to modern technologies; even when these are locally developed, they are more complex to use, repair, and maintain. They are also expensive and tend to rely on imported components and non-renewable sources of energy (Fellows *et al.*, 1993).

The Image of Packages

Packaging and packages have a bad image, in some part thanks to environmental activists. Probably glass and metal are the least harmful thought of and plastics and composite materials the most.

The bad image of packaging, however, does not translate into consumer hostility at point of sale. Products and not packages are bought. Packages are not noticed during purchase/transport/use of the product, but the minute the product is consumed, and the package had fulfilled its function, it turned waste. At that minute it is already seen as an environmental burden, something what is "polluting the environment".

It is convenient to fill the shopping trolley with goods, then stuff it all into plastic bags, place them into the boot of the car, thus get them all safe home. One seldom realises that it is exactly thanks to packaging that it is possible. Indeed it is more annoying to handle goods with no, or just light package. It is much easier is to carry home goods in rigid plastic bottle, or in aseptic box. But the very same plastic bottle or aseptic box is seen as nuisance, environmentally unfriendly when it ends up in the waste-basket. Then suddenly one's environmental consciousness arise, and thinks of what could the industry, retailers or government do about "all those" packages.

In the same time, the same consumer buys French apples, kiwi from New Zealand, orange from Israel, drinks Colombian coffee and Chinese tea, seasons the food with spices from India or Africa, uses car, takes baths, uses dishwashers, washing machines, all of what means tremendous amount of (drinking) water consumption. At public places one uses paper towels. The thing, however, what, through the media, seems the most conceived as a harm done to the environment for consuming resources, is the plastic bottle, and aseptic box in the waste basket. One may feel somewhat relieved to find containers for collecting them separately. During the pilgrimage to these containers the citizen may feel that he has done something good for the environment's sake, and does not necessarily want to know whether it is really good or not. One would probably argue that food and drinks, toiletry and cleaning products are needed. To get them, however, we need packaging as well.

We cannot discuss food packaging without discussing the effects of packaging waste on the environment. Clearly, recycling is a sound approach. However, the problem often lies in feasibility of collection, separation and purification of the consumerís disposed food packages. This mode of recycling is called post-consumer recycling. While it offers a logistic challenge, recycling is gaining in popularity, and the packaging industry is cooperating in that effort. Aluminum cans are the most recycled container at this time. Plastic recycling is increasing, yet most plastic is recycled during manufacturing of the containers; not as post-consumer recycling. For example, trimmings from plastic bottles are reground and reprocessed into new ones.

The plastics industry is helping to facilitate consumer recycling by identifying the type of plastic from which the container is made. A number from 1 to 7 is placed within the recycling logo on the containerís bottom. For example, 1 refers to PET (Polyethylene Terephthalate), the plastic used for the large 2 liter soft drink bottles. Plastics have the advantage of being light. This helps to conserve fuel during transport and also reduces the amount of package waste.

There are many interesting packaging concepts being explored by the industry to keep up with the changing life style of the consumer and new technologies. Many professionals are involved in designing and manufacturing the modern package. Today's package is designed with the consumer's safety and convenience in mind. Examples are microwaveable popcorn packages, squeezable ketchup bottles and the tamper-proof milk bottle cap.

The Influence of Packaging on the Environment

Packaging in the modern world has a considerable impact on the environment. Food packaging makes up two fifths of the household waste (Senaner *et al.*, 1991). Life-cycle assessment for new packaging materials, such as nanobased and bio-based materials, is essential for their success. This is a comprehensive analysis of materials from production to disposal to determine their environmental impact. It involves evaluation of the materials and energy usage in product manufacture and use and evaluation of the type and amount of waste generated.

Uniform criteria or a universal model with standard inputs and outputs is needed to evaluate packaging materials. Analysis of the performance and sustainability of new materials, especially biobased materials, is also required to document their advantages over conventional petroleum-based materials. Investigations into the compatibility of new materials with food products are important.

Environmental Concerns that Impact Material Choices

Whether considering a full life cycle analysis or the packaging conversion process alone, polyethylene and polypropylene are materials that favourably address environmental concerns such as energy consumption, climate protection, and water conservation, compared to competing and traditional packaging materials.

Polyethylene and polypropylene intrinsically offer low densities which mean low packaging unit weight. This results in an important reduction of material used and minimises waste at end of life. The materials also significantly lower fuel consumption for transport and contributes to lowering our CO_2 emissions. In addition, advanced polyolefins are recyclable or can be turned back into energy to substitute fossil fuels in clean energy recovery processes.

Reducing the Environmental Impact of Packaging Materials

The food companies are also concerned that the products are designed to be eco-friendly. In particular, efforts are underway to reduce the weight of packaging materials (which turn into waste after consumption), adopt materials with less environmental impact, and collect and recycle used containers. In Japan, with the Containers and Packaging Recycling Law instituted in 1995, municipalities began to separate waste and supermarkets placed several collection boxes at their storefronts to recycle packaging materials. Public efforts to recycle packaging materials are thus, very important for the environmental health.

Review and Change of Food Packaging Materials

Food packaging materials play a primary role in protecting the contents from ultraviolet rays, germs in the air, shocks during transportation, etc., while they turn into waste after consumption. Of non-industrial waste comprised primarily of household garbage, waste packaging materials account for some 60 per cent of the total volume, and 20-30 per cent of the total weight, according to some estimates. Besides, they need to be collected and treated separately as they are usually made of several materials, a factor that makes the recycling of packaging materials rather difficult.

For this reason, the companies have begun to change the packaging materials for frozen foods in February 2002, from composite materials (aluminized polypropylene or nylon) to polypropylene. The design and strength of inner trays were also reviewed for weight reduction purposes. All these efforts are designed to simplify the recycling of waste packaging materials and reduce their environmental impact (new packaging materials, when incinerated, produce less CO_2 and hazardous gases). The food companies are now committed to providing customers with products designed to be eco-friendly, thereby fulfilling corporate social responsibility.

Recycle Plaza JB (located in Saitama City, Japan), an intermediate treatment plant for used beverage containers, has been in operation since May 2003. This plant was jointly established by JT and its affiliate, Japan Beverage, a vending machine operator. These beverage vending machines are installed in various locations such as office buildings, factories and along roadsides, the disposal of used containers had always been a major concern to the company. According to Mr. Toru Sunaga, manager of the environmental promotion division, Japan Beverage Inc., they felt that they were responsible for taking care of the entire process, from sales of products to collection and recycling of used containers – which resulted in the establishment of Recycle Plaza JB, an intermediate treatment plant for used containers.

Recycle Plaza JB is capable of treating 64 tons of used containers a day, which include aluminum cans, steel cans, paper cups, and brick packs collected by Japan Beverage's 38 sales offices in the Kanto area. It is the first of its kind managed by a vending machine operator, recycling a variety of beverage containers. He reported that they are, and will be, promoting recycling efforts on a company-wide basis, based on the awareness that paying attention to the environment is a critical theme that should be addressed not only by companies but also by each individual.

Going Green on Fast Food Packaging

With the hectic lifestyle of today, more and more people have less and less time to walk around and take in the scenery, let alone cook their own food. A lot of people rely on the existence of a burger joint or some other variation of fast food chain that can be located around every turn of each street corner. Upon receiving and paying for their ordered food, few people take the time to notice how their food has been packaged.

Fast food packaging is something that has been talked about and debated about by various environmentalists, primarily due to the type of materials that are being used as utensils and containers. Fast food joints usually make use of plastic spoons and forks, Styrofoam or polystyrene thermal insulation food containers, plastic or Styrofoam cups, plastic straws and stirrers, greaseproof and sulphite bags and burger wraps. For potato fries, cardboard containers or paper containers are used.

As can already be observed, fast food packaging consists of disposable items. Some environmentalists would argue that these disposable items, particularly the ones made of polystyrene thermal insulation material can be harmful to the earth if not properly disposed or recycled.

Although plastic is recyclable, most of these materials usually end up in dumps and never really decays creating more garbage. Fast food conglomerates argue that the use of disposable products is environmentally friendly or at the very least acceptable since less water is used. With these disposable fast food packagings, dishwashers would not need to be working around the clock saving more energy and water.

There are some fast food joints these days that prefer using recycled paperbags for take aways or take-out meals. This is a fairly good move since these items are biodegradable. Cardboard boxes are also being used by some fast food chains as opposed to using Styrofoam containers when serving their burgers. Another environment-friendly alternative implemented by top fastfood chains include the use of unbleached paperbags as opposed to bleached paperbags or plastic for take away food.

The progress of making fast food packaging more environment-friendly is rather sluggish, and the great amount of polystyrene wastes that are accumulating is increasingly becoming alarming because of the pollution they produce. There are some countries that encourage their citizens and the fast food chains to do their share of saving the environment by proper trash disposal and recycling. If more fast food chains and consumers would care about efficient and environment-friendly fast food packaging, then perhaps the progress of revolutionizing the packaging system would become faster.

Eco-friendly Consumer Choices

There aren't perfect solutions, short of buying foods only in their whole, natural form.

Given the culture we live in, we inevitably purchase packaged and throw-away items.

Yet, as non-renewable resources dwindle and technology advances, new possibilities are emerging.

PCC merchandisers have considered several novel ideas for containers and packaging with an eye toward more sustainable options. They're looking at containers made from palm fiber, a waste product of the palm oil process, and they're investigating new "food films" made from non-GE eucalyptus pulp as an alternative to plastic wrap. Both products are recyclable, biodegradable and can compost in own backyard. In the meantime, for shoppers who buy throwaway products, PCC offers bioplastic utensils made from non-GE corn and paper products made from 100 per cent recycled paper and no chlorine. When reviewing new items, merchandisers always give preference to products with less packaging.

Retailers, however, can't alter the marketplace alone. Consumers need to take responsibility, too, and support more sustainable solutions:

1. Bring your own glass or reusable containers for daily foods.
2. Re-use paper or plastic bags for bulk flour, nuts, etc.
3. Use long-lasting canvas bags for packing groceries to take home.
4. For fresh produce, don't use a bag since nature provides a natural skin as a "wrapper" and all produce can be packed safely at the register for transport home.
5. Call ahead and ask for butcher paper instead of plastic wrap for meat.
6. Choose products without packaging or reduced packaging, such as berries in fiberboard instead of plastic boxes.

The old maxim "Reduce, reuse, recycle" still holds. Yet recycling and reusing a product really starts with buying products that can be re-used. This is called "precycling." After all, when we throw something away, there is no away. To be a sustainable retailer, we need the consumers to make sustainable choices.

The food packaging industry is committed to developing products and implementing practices that can contribute to a cleaner, safer and better world.

Sustainable Packaging to Protect the Environment

Sustainability of packaging means:

1. To strengthen or support
2. To keep something going
3. Over time or continuously

It is the concept of using technology for meeting the needs of 'today'without in any way compromising the interests of future generations or the 'tomorrow'. The requirements of sustainable technology are:

1. Must not impose any burdens on the environment or material resources, renewable or non-renewable.
2. Must not cause deterioration in their quality or availability.

To address the above mentioned concerns, we should look into the issues to achieve the objective of sustainable packaging as it has the following advantages:

1. It preserves natural resources like non-renewable including fossil fuels and minerals. Packaging material should prevent depletion of non- renewable resources.
2. Sustainable packaging is beneficial, safe and healthy for individuals and communities throughout its life cycle. It meets the market criteria for performance and cost.
3. It can save other scarce resources like energy. It should be physically designed in such a way as to optimise materials and energy. Packaging material should conserve energy efficiency. It should be sourced, manufactured, transported and recycled using renewable energy.

4. Packaging should be eco-friendly. It should prevent pollution of air and water, global warming, and not have an adverse effect on forest cover. The material to be used should be such that it preserves the quality of environment. It maximises the use of renewable recycled source materials and is manufactured using clean production technologies and best practices. It is made from materials healthy in all probable end of life scenarios.

5. The packaging material should not generate but reduce waste. Use renewable and recyclable materials. Use cleaner and greener processes and should avoid or reduce greenable emissions. It should be effectively recovered and utilised in biological and/or industrial cradle to cradle cycles.

Chapter 18

The Vision for Future Packaging

Lessons in Packaging Learned from Nature

Perfect packaging exists in nature - examples include the banana and the egg, together with the many smart materials and systems that control plant and biological functions. Learning from Nature to solve engineering problems (biomimetics) is not new, although the term itself is. Animals and plants have evolved many successful structural and functional mechanisms and increasing our study of biological materials and systems will almost certainly yield promising engineering concepts applicable to smart packaging. Packaging in the future could indeed be 'smart by name, smart by Nature!'

The vision of the future of packaging, according to the recently published Foresight report 'Materials: Shaping Our Society', is one in which the package will increasingly operate as a smart system incorporating both smart and conventional materials, adding value and benefits across the packaging supply chain. For smart materials to be adopted in packaging, they need to be inexpensive relative to the value of the product, reliable, accurate, reproducible in their range of operation, and environmentally benign and food contact safe.

Smartness in Packaging

'Smartness' in packaging is a broad term that covers a number of functionalities, depending on the product being packaged, including food, beverage, pharmaceutical, household products etc. Examples of current and future functional 'smartness' would be in packages that:

1. Retain integrity and actively prevent food spoilage (shelf-life)
2. Enhance product attributes (*e.g.* look, taste, flavour, aroma etc)
3. Respond actively to changes in product or package environment
4. Communicate product information, product history or condition to user
5. Assist with opening and indicate seal Integrity
6. Confirm product authenticity, and act to counter theft.

Smart Packaging and Activated Packaging

There is an important distinction between package functions that are smart/intelligent, and those that become active in response to a triggering event, for example, filling, exposure to UV, release of pressure etc and then continue until the process is exhausted. Some smart packaging already exists commercially and many other active and intelligent packaging concepts are under development (Table 18.1). A good example of active packaging is the highly successful foam-producing 'widget' in a metal can of beer. Another is the oxygen scavenging MAR technology patented by CMB Packaging Technology (now Crown Cork & Seal).

RFID Tags in Packaging: Smart Packaging

Packaging technology could possibly realize its most influential innovations through smart packaging. Smart packaging encompasses a wide range of technologies including films that indicate product freshness, as well as security inks used to protect against counterfeiting in the food and pharmaceutical sector.

Perhaps the most significant technological advancement currently evolving is radio frequency identification (RFID). The use of RFID tags in packaging is increasing for all end-markets; however, a majority of current activity can be seen in the food and beverage and pharmaceutical sectors. RFID will play a crucial role in packaging for the foreseeable future.

Table 18.1: Smart Packaging Under Development

Active	Intelligent
– Oxygen scavenging	– Time-temperature history
– Anti-microbial	– Microbial growth indicators
– Ethylene scavenging	– Light protection (photochromic)
– Heating/cooling	– Physical shock indicators
– Odour and flavour absorbing/releasing	– Leakage, microbial spoilage indicating
– Moisture absorbing	

RFID tags could be used on food packaging to perform relatively straightforward tasks, such as allowing cashiers in supermarkets to tally all of a customer's purchases at once or alerting consumers if products have reached their expiration dates. RFID tags are controversial because they can transmit information even after a product leaves the supermarket. Privacy advocates are concerned that marketers will have even greater access to data on consumer behavior. Wal-Mart in the US and TESCO in the UK have already tested RFID tagging on some products in some stores. The tagging of food packages will mean that food can be monitored from farm to fork during processing, while in transit, in restaurants or on supermarket shelves and eventually, even after the consumer buys it. Coupled with nanosensors, those same packages can be monitored for pathogens, temperature changes, leakages etc. (Scott, 2002).

Wal-Mart is two years into the world's most ambitious effort to implement RFID technology and is making significant headway with its Next Generation logistics initiative. The company kicked-off their RFID program in 2004 with suppliers that included Gillette, Hewlett-Packard, Johnson & Johnson, Kimberly-Clark, Kraft Foods, Nestlé Purina PetCare, Procter & Gamble, and Unilever, and has already begun to realize a return on its investment with improved sales of promotional display items and a sizable reduction in product stock-outs.

It is estimated that the market for smart technology in packaging should grow to nearly $5 billion by 2011 and reach $14 billion by 2014. As electronic printing technologies improve and smart packaging technologies become more economically viable, interoperable RFID products will allow for improved inventory management, logistics, and supply chain management on a global scale.

Smart Packaging Concepts for Pharmaceuticals

Smart packaging concepts that improve case of use could include 'dial-a-dose' smart caps and closures that allow the safe dispensing of exact controlled quantities of product, *e.g.* pharmaceuticals, cleaning materials, and other potentially hazardous materials. Already a prescription drug bottle with bottle cap alarm is available - it beeps to alert users when it is time to take the medication, and it displays how many times the bottle has been opened and the intervals between openings. The bottle can be connected via a modem to the healthcare centre for the automatic transmission of drug usage and, if necessary, provide feedback to the patient if not in compliance. Eventually, programmed skin patches using smart gels that rely on changes in skin properties to trigger drug delivery could replace conventional pill-taking medication.

How Activated Packaging Systems Work

This consists of a matrix polymer, such as PET, an oxygen scavenging/absorbing component and a catalyst. The oxygen-scavenging component is a nylon polymer (MXD6) melt blended with the PET at around the 5 per cent level. The catalyst is a cobalt salt added at a low concentration (less than 200ppm) that triggers the oxidation of the MXD6. The OXBAR system remains active for periods of up to two years providing protection to oxygen sensitive products such as beer, wine, fruit juice and mayonnaise throughout their shelf-lives. Active food packaging systems using oxygen scavenging and anti-microbial technologies (*e.g.* sorbate-releasing LDPE film for cheese) have the potential to extend the shelf-life of perishable foods while at the same time improving their quality by reducing the need for additives and preservatives.

How Intelligent Packaging Works

In 'intelligent' packaging, the package function switches on and off in response to changing external/internal conditions and can include a communication to the customer or end user as to the status of the product. A simple definition of intelligent packaging is 'packaging which senses and informs', and nowhere does this generate a more potent vision than within the smart home of the future (Butler, 2001).

Intelligent Tamper-Proof Packaging

Knowing whether a package has been tampered with is equally important to consumers. Tamper evidence technologies that cannot easily be replicated, *e.g.* based on optically variable films or gas sensing dyes, involving irreversible colour changes, will become more widespread and cost-effective for disposable packaging of commodity items. Piezoelectric polymeric materials might be incorporated into package construction so that the package changes colour at a certain stress threshold. In this way, a 'self bruising' closure on a bottle or jar might indicate that attempts had been made to open it.

Easier to Open Packaging

Easier to open packaging will be a paramount feature of future packaging. A recent DTI survey showed that in the UK in 1997, 90,964 accidents requiring hospital treatment were packaging related. The focus will be on better design (size, shape, etc.) and the optimum use of materials, to produce easy to open packages consistent with the strength capabilities of an ageing population. Developments in low peel-force adhesives and structures, even smart packages that are self-opening, *e.g.* based on shape memory alloys (the metal 'rubber' band), can be envisaged.

Quality Assurance Using Intelligent Labels

Another important need is for consumer security assurance, particularly for perishable food products. The question as to whether, for example, a chilled ready-meal is safe to use or consume is currently answered by 'best by' date stamping. However, this does not take into account whether the product has inadvertently been exposed to elevated temperatures during storage or transportation. In the future, microbial growth and temperature-time visual indicators based on physical, chemical or enzymatic activity in the food will give a clear, accurate and unambiguous indication of product quality, safety and shelf-life condition. As an example, COX Technologies has developed a colour indicating tag that is attached as a small adhesive label to the outside of packaging film, which monitors the freshness of seafood products. A barb on the backside of the tag penetrates the packaging film and allows the passage of volatile amines, generated by spoilage of the seafood. These are wicked passed a chemical sensor that turns FreshTag progressively bright pink as the seafood ages.

Intelligent Packaging for Fresh Fruit and Vegetables

Fresh-cut produce continues to be one of the fastest growing segments of food retailing and while conventional film packaging is suitable for lettuce and prepared salads, it cannot cope with the high respiration rates of pre-cut vegetables and fruit, leading to early product deterioration. In the USA, novel breatheable polymer films are already in commercial use for fresh-cut vegetables and fruit. Landec Corporation supplies Intellipac packaging films that are acrylic side-chain crystallisable polymers tailored to change phase reversibly at various temperatures from 0-68°C. As the side-chain components melt, gas permeation increases dramatically, and by further tailoring the package and materials of construction, it is possible to by tune the carbon dioxide to oxygen permeation ratios for particular products. The final package is 'smart' because it automatically regulates oxygen ingress and carbon dioxide egress by transpiration according to the prevailing temperature. In this way, an optimum atmosphere is maintained around the product during storage and distribution, extending freshness and allowing shipping of higher quality products to the consumer.

Self-Heating and Self-Chilling Packaging

Improved convenience is a value-added function that customers are likely to pay extra for as lifestyles change. Self-heating packages, for soup and coffee, for example, and self-cooling containers for beer and soft drinks have been under active development for more than a decade, but have yet to achieve commercial status. However, Crown Cork & Seal is pioneering the development of a self-chilling beverage can in conjunction with Tempra Technologies and development is nearing completion. The Crown/Tempra technology uses the latent heat of evaporating water to produce the cooling effect. The water is bound in a gel layer coating a separate container within the beverage can, and is in close thermal contact with the beverage. The consumer twists the base of the can to open a valve, exposing the water to the desiccant held in a separate, evacuated external chamber. This initiates evaporation of the water at room temperature. The unit has been designed to meet a target specification set by major beverage customers cooling 300ml of beverage in a 355ml can by 16.7°C in three minutes. This performance level has been achieved in laboratory tests and working samples are currently undergoing focus group trials with customers.

Thermochromic Labelling

Give a self-heating or self-cooling container a sensor to tell the consumer it is at the correct temperature and the package becomes 'smart' (such packaging is currently under development). The most common use a thermochromic ink dot to indicate the product is at the correct serving temperature following refrigeration or microwave heating. Plastic containers of pouring syrup for pancakes can be purchased in the USA that are labelled with a thermochromic ink dot to indicate that the syrup is at the right temperature following microwave heating. Similar examples can be found on supermarket shelves with beer bottle labels that incorporate thermochromic-based designs to inform the consumer when a refrigerated beer is cold enough to drink.

Factors that Aid the Growth of Intelligent Packaging

Consumer and societal factors are likely to drive the adoption of smart packaging in the future. The growing need for information on packaging will mean there has to be a step change in providing this information. Consumers increasingly need to know what ingredients or components are in the product and how the product should be stored and used. Intelligent labelling and printing, for example, will be capable of communicating directly to the customer via thin film devices providing sound and visual information, either in response to touch, motion or some other means of scanning or activation. Voice-activated safety and disposal instructions contained on household and pharmaceutical products will be used to tell the consumer how they should be disposed of after consumption - information that can be directly used in the recycling industry to help sort packaging materials from waste streams. Drug delivery systems in smart packaging will be programmed to communicate patient information back to healthcare centres.

Possible Concerns over Intelligent Packaging

When it comes to the environment, consumer attitudes towards packaging are generally confused and contradictory. There are increasing concerns about the amount of waste created by packaging, but the growth of more elaborate and attractive packaging is being driven by consumers to fuel our desire for convenience and feed our lifestyle choices. Ultimately, future consumers could react negatively to the perception of increased waste and lack of recyclability of disposable smart packages. The perception of extra cost and complexity, and the possible mistrust/confusion of technology - for example, if there is both a date stamp and a visual indicator on a food pack, which does the customer take note of - are further factors that could slow widespread market introduction of smart packages. The overall acceptance barriers to smart packaging can be summed up as:

1. Extra cost - can it be absorbed/passed on to consumer?
2. Unreliability of indicating devices showing either food to be safe when it is not (potential liability?) or food to be unsafe when it is (increased spoilage stock loss)
3. Food safety and regulatory issues - *e.g.* possible migration issues of complex packaging materials into product
4. Recycling features and environmental regulations.

What Areas will Benefit from Intelligent Packaging First?

Cost issues will probably mean that early adopters of smart packaging are likely to be in non-commodity products, *e.g.* pharmaceuticals, health and beauty, and packaging that plays a part in lifestyle and leisure activities. A further consideration is the need for education to reassure the consumer of package safety, and ensure against incorrect operation and mistrust of smart technology. The successful adoption of smart packaging concepts in the future must create advantages for the whole of the supply chain.

Bibliography

Ahmed E. and Bird, K. 2007. Nanoscale particles to block UV light in plastic packaging. *Nanowerk LLC* Report.

Anonymous (2004).Down on the Farm: the Impact of Nano-Scale Technologies on Food and Agriculture, ETC Group Report.

Asadi, G. and Mousavi, M. Application of Nanotechnology in Food Packaging.

Brody, A.L. (1989). Modified atmosphere packaging of Seafoods. In: *Controlled/Modified Atmosphere Vacuum Packaging of Foods*, A.L. Brody (Ed.), Food and Nutrition Press, Inc., Trumbull, Connecticut, Chap. 4.

Brooks, J. (1929). *Biochem. J.*, **23**: 1391.

Brooks, J. (1931). *Proc. Royal Soc., London, Ser. B.*, **109**: 35.

Brooks, J. (1937). *Proc. Royal Soc., London, Ser. B.*, **123**: 368.

Brooks, J. (1938). *Food Res.*, **3**: 75.

Butler, O.D., L.J. Bratzler and W.L. Mallman (1953). *Food Technol.*, **7**: 397.

Butler, P. 2001. Smart Packaging - Intelligent Packaging for Food, Beverages, Pharmaceuticals and Household Products. Materials World, Vol. 9, No. 3 pp. 11-13.

Clark, D.S. and C.P. Lentz (1969). *Can. Inst. Food Sci. Technol. J.*, **2**: 72.

Clark, D.S. and C.P. Lentz (1972). *Can. Inst. Food Sci: Technol. J.*, **5**: 175.

Clark, D.S. and C.P. Lentz (1973). *Can. Inst. Food Sci. Technol. J.*, **6**: 194.

Clark, D.S., C.P. Lentz and L.A. Roth (1976). *Can. Inst. Food Sci. Technol. J.*, **11**: 363.

Cole, A.B. (1986). Reciprocal Meat Conference Proc., **39**: 106.

Dainty, R.H., B.G. Shaw and T.A. Roberts (1983). Microbial and chemical changes in chill-stored red meats. In: *Food Microbiology: Advances and Prospects*, T.A. Roberts and F.A. Skinner (Eds.), Academic Press, London, England, p. 151.

Daniels, J.A., R. Krishnamurth and S.S.H. Rizvi (1985). *J. Food Protect.*, **48**: 532 (1985).

Dawson, L.E. (1987). Packaging of processed poultry. In: *The Microbiology of Poultry Meat Products*, F.E. Cunningham and N.A. Cox (Eds.), Academic Press, Inc., London, Chap. 7.

Denton, J.H. and F.A. Gardner (1981). *J. Food Sci.*, **47**: 214.

Donald, A. 2004. Food for thought. Nature Materials. 3: 578-581.

Enfors, S.-O. and G. Molin (1980). *Can. J. Microbiol.*, **27**: 15.

Enfors, S.-O. and G. Molin (1981). *J. Appl. Bacteriol.*, **48**: 409.

Enfors, S.-O. and G. Molin (1981). The effect of different gases on the activity of microorganisms. In: *Psychrotrophic Microorganisms in Spoilage and Pathogenicity*, T.A. Roberts, G. Hobbs, J.H.B. Christian and N. Skovgaard (Eds.), Academic Press, London, England, Chap. 34.

Farber, J.M. (1991). *J. Food Protect.*, **54**: 58.

Finne, G. (1981). Principles of controlled atmosphere extension of food shelf life. In: *Proc. Int. Conf. Control. Allnos., CAP'84*, Chicago, Illinois, p. 37.

Gardner, G.A. (1983). Microbial spoilage of cured meats. In: *Food Microbiology: Advances and Prospects*, T.A. Roberts and F.A. Skinner (Eds.), Academic Press, London, England, p. 179.

Genigeorgis, C. (1985). *Int. J. Food Microbiol.*, **1**: 237.

Georgala, D. and C.M. Davidson (1970). *British Patent*, **1**: 199, 998.

George, P. and C.J. Stratmann (1952). *Biochem. J.*, **51**: 418.

Gibbs, P.A. and J.T. Patterson (1977). Some preliminary findings on the microbiology of vacuum- and gas-packaged chicken portions. In: *The Quality of Poultry Meat*, S. Scholtyssek (Ed.), Universitat Hohenheim, Stuttgart, p. 198.

Gill, C.O. (1982). Microbial interaction with meats. In: *Meat Microbiology*, M.H. Brown (Ed.), Applied Science Publishers Ltd., London, England, Chap. 7.

Gill, C.O. (1988). *Meat Sci.*, **22**: 65.

Gill, C.O. (1989). *Brit. Food J.*, **91(7)**: 11.

Gill, C.O. (1990). *Food Control J.*, **74**.

Gill, C.O. and K.H. Tan (1980). *Appl. Environ. Microbiol.*, **39**: 317.

Gill, C.O. and N. Penney (1985). *Meat Sci.*, **14**: 43.

Gill, C.O. and N. Penney (1986). *Meat Sci.*, **18**: 41.

Gill, C.O., C.L. Harrison and N. Penney (1990). *Int. J. Food Micro.*, **11**: 151.

Grau, F.H. (1983). *J. Food Sci.*, **48**: 326.

Harte, B.R. (1987). Packaging of restructured meats. In: *Advances in Meat Research*, A.M. Pearson and T.R. Dutson (Eds.), Elsevier Applied Science Publishers, Essex, England, Chap. 13.

Hauschild, A.H.W. and B. Simonsen (1985). *J. Food Protect.*, **48**: 997.

Hayakawa, K., Y.S. Henig and S.G. Gilbert (1975). *J. Food Sci.*, **40**: 186.

Henig, Y.S. and S.G. Gilbert (1975). *J. Food Sci.*, **40**: 1033.

Herner, R.C. (1987). High CO_2 effects on plant organs. In: *Postharvest Physiology of Vegetables*, J. Weichmann (Ed.), Marcel Dekker, Inc., New York, Chap. 11.

Hicks, A. 2002. Minimum Packaging Technology for Processed Foods: Environmental Considerations. AU J.T. 6(2): 89-94.

Hintlian, C.B. and J.H. Hotchkiss (1986). *Food Technol.*, **40(12)**: 70.

Hirose, K., B.R. Harte, J.R. Giacin, J. Miltz and C. Stine. (1988). Sorption of d-limonene by sealant films and effect on mechanical properties. In: *Food and Packaging Interactions*, J.H. Hotchklss (Ed.), ACS Symp. Ser. #365, American Chemical Society, Washington, D.C., Chap. 3.

Hobbs, G. (1983). Microbial spoilage of fish. In: *Food Microbiology: Advances and Prospects*, T.A. Roberts and F.A. Skinner (Eds.), Academic Press, London, England, p. 217.

Holdsworth, S.D. (1983). *The Preservation of Fruit and Vegetable Food Products*. The Macmillan Press Ltd, London, England.

Huxsoll, C.C. and H.R. Bolin (1989). *Food Technol.*, **43**: 124.

Isenberg, M.F.R. (1979). *Hort. Rev.*, **1**: 337.

Jackson, J.M. and B.M. Shinn (1979). *Fundamentals of Food Canning Technology*. AVI Publishing Company, Inc., Westport, Connecticut.

Jurin, V. and M. Karel (1963). *Food Technol.*, **17(6)**: 104.

Kader, A.A. (1980). *Food Technol.*, **34(3)**: 51.

Kader, A.A. (1986). *Food Technol.*, **40(5)**: 99.

Kader, A.A. (1987). Respiration and gas exchange of vegetables. In: *Postharvest Physiology of Vegetables*, J. Weichmann (Ed.), Marcel Dekker, Inc., New York and Basel, Chap. 3.

Kader, A.A., D. Zagory and E.L. Kerbel (1989). *CRC Crit. Rev. Food Sci. Nutr.*, **28**: 1.

Kader. A.A. (1985). Postharvest and biology. In: *Postharvest Technology of Horticultural Crops*. A.A. Kader, R.F. Kasmire, F.G. Mitchell, M.S. Reid, N.F. Somer and J.F. Thompson, Cooperative Extension University of California Division of Agriculture and Natural Resources Special Publication 3311, Chap. 2.

Karel, M. and J. Go (1964). *Modern Pack.*, **37(6)**: 123.

Karel. M. (1975). Protective packaging of foods. In: *Principles of Food Science Part II: Physical Principles of Food Preservation*, M. Karel. O.R. Fennema and D.B. Lund, Marcel Dekker, Inc., New York, Chap. 12.

Kawada, K. (1982). Use of polymeric films to extend postharvest life and improve marketability of fruits and vegetables-UNIPACK: individually wrapped storage of tomatoes, oriental persimmons and grapefruit. In *Proceedings of the Third National Controlled Atmosphere Research Conference on Controlled Atmospheres for Storage and Transport of Perishable Agricultural Commodities*, D.G. Richardson and M. Meheriuk (Eds.), Timber Press, Beaverton, Oregon, p. 87.

King, A.D and H.R. Bolin (1989). *Food Technol.*, **43**: 132.

Kramer, A., T. Solomos, F. Wheaton, A. Puri, S. Sirivichaya, Y. Lotem, M. Fowke and L. Ehrman (1980). *Food Technol.*, **34**: 65.

Labuza, T.P. (1984). *Moisture Sorption: Practical Aspects of Isotherm Measurement and Use.* American Association of Cereal Chemists, St Paul. Minnesota.

Lakin, W.D. (1987). *J. Packag. Tech.*, **1**: 82.

Lee, D.S., P.E. Haggar, J. Lee and K.L. Yam (1991). *J. Food Sci.*, **56**: 1580.

Lee, J.J.L. (1987). The design of controlled or modified packaging systems for fresh produce. In: *Food Product: Package Compatibility,* J.I. Gray, B.R. Harte and J. Miltz (Eds.), Technomic Publishing Co., Inc., Pennsylvania, p. 157.

Lipton, W.J. (1975). Controlled atmospheres for fresh vegetables and fruits why and when. In: *Postharvest Biology and Handling of Fruits and Vegetables,* N.F. Haard and D.K. Salunke (Eds.), AVI Publishing Co. Inc, Wesport, Connecticut, p. 130.

Lougheed, E.C. (1987). *Hort Sci.*, **22**: 791.

Mannapperuma, J.D., D. Zagory, R.P. Singh and A.A. Kader (1989). Design of polyme packages for modified atmosphere storage of fresh produce. *Proceedings 5th Controlled Atmosphere Research Conference,* Vol. 2, Wenatchee, Washington, p. 225.

Mannheim, C.H., J. Miltz and A.J. Letzter (1987). *J. Food Sci.*, **52**: 737.

Metlitskii, L. V., E.G., Sal'Kova, N.L. Volkind, V.I. Bondarev and V. Yanyuk (1983). *Controlled Atmosphere Storage of Fruits.* Amerind Publishing Co. Pvt. Ltd., New Delhi, India, Chap. VI.

Moraru, C., Panchapakesan,C., Huang Q. and Takhistov, P., 2003. Nanotechnology: A new frontier in food science. Food Technol. 57: 25-27.

Myers, R.A. (1989). *Food Technol.*, **43**: 129.

Nelson, P.E. and D.K. Tressler (1980). *Fruit and Vegetable Juice Processing Technology,* 3rd edn., AVI Publishing Co. Inc., Westport, Connecticut.

Pantastico, E.B., T.K. Chattopadhyay and Subramanyam, H. (1975). Storage and commercial operations. In: *Postharvest Physiology,*

Handling and Utilization of Tropical and Subtropical Fruits and Vegetables, E.B. Pantastico (Ed.), AVI Publishing Co. Ltd., Connecticut, Chap. 16.

Peleg, K. (1966). Automatic wrapping and packaging of citrus. M.Sc. Thesis, Technion Haifa, Israel.

Peleg, K. (1985). Retail produce packaging. In: *Produce Handling, Packaging and Distribution*, AVI Publishing Co. Inc., Wesport, Connecticut, Chap. 11.

Powrie, W.D. (1988). Optimization of parameters to prolong retention of prepackaged fruit. In: *Proceedings of the Fourth International Conference on Controlled Atmosphere Packaging CAP'88*, Scotland Business Research, Inc., New Jersey, p. 245.

Powrie, W.D. and Skura, B.J. (1991). Modified atmosphere packaging of fruits and vegetables. In: *Modified Atmosphere Packaging of Food*, B. Ooraikul and M.E. Stiles (Eds.), Ellis Horwood Ltd., Chichester, England, Chap. 7.

Prince, T.A. (1989). Modified atmosphere packaging of horticultural commodities. In: *Controlled/Modified Atmosphere/Vacuum Packaging of Foods*, A.L. Brody (Ed.), Food and Nutrition Press, Trumbull, Connecticut, Chap. 5.

Reid, M.S. (1985). Ethylene in postharvest technology. In: *Postharvest Technology of Horticultural Crops*, A.A. Kader, R.F. Kasmire, F.G. Mitchell, M.S. Reid, N.F. Somer and J.F. Thompson, Cooperative Extension University of California Division of Agriculture and Natural Resources Special Publication 3311, Chap. 13.

Reitman, N. and Doppelt, B. 2007. Packaging and Plastics: What's a Consumer to Do? PCC Natural Markets' Newsletter: The Sound Consumer.

Risse, L.A., and W.R. Miller (1986). *J. Plastic Film & Sheeting*, **2**: 163.

Salunke, D.K. and B.B. Desai (1984). *Postharvest Biotechnology of Vegetables*, CRC Press, Inc., Florida, Vols. I & II.

Scott, A.(2002). BASF takes big steps in small tech, focusing on nanomaterials. Chemical Week.164,1-45.

Smock, R.M. (1979). *Hort. Rev.*, **1**: 301.

Tolle, W.E. (1971). Variables affecting film permeability requirements for modified-atmosphere storage of apples. *U.S. Dept. Agr. Tech. Bull.*, No. 1422.

Troller, J.A. and J.H.B. Christian (1978). *Water Activity and Foods,* Academic Press Inc., New York.

Veeraju, P. and M. Karel (1966). *Mod. Pack.,* **40(2)**: 168.

Weichmann, J. (1987). Low oxygen effects. In: *Postharvest Physiology of Vegetables,* J., Weichmann (Ed.), Marcel Dekker, Inc., New York and Basel, Chap. 10.

Yang, C.C. and M.S. Chinnan (1988). *J. Food Sci.,* **53**: 869.

Zagory, D., and A.A. Kader (1988). *Food Technol.,* **42(9)**: 70.

Index